VEHICLE MAINTENANCE AND REPAIR

LEVEL 3

T0175747

David Hobson, Patrick Hamilton and Graham Baker

Edited by Roy Brooks

CENGAGE
Learning

Australia • Brazil • Japan • Korea • Mexico • Singapore • Spain • United Kingdom • United States

CENGAGE
Learning·

Vehicle Maintenance and Repair Level 3

David Hobson, Patrick Hamilton and Graham Baker

Edited by Roy Brooks

Development Editor: Catharine Esmat

Senior Project Editor: Alison Burt

Senior Manufacturing Buyer: Eyvett Davis

Typesetter: MPS Limited

Cover design: HCT Creative

Text design: Design Deluxe, Bath

For product information and technology assistance, contact **emea.info@cengage.com**.

For permission to use material from this text or product, and for permission queries, email **emea.permissions@cengage.com**.

British Library Cataloguing-in-Publication Data
A catalogue record for this book is available from the British Library.

ISBN 13: 978-1-4080-7754-2

Cengage Learning EMEA
Cheriton House, North Way, Andover, Hampshire SP10 5BE United Kingdom

Cengage Learning products are represented in Canada by Nelson Education Ltd.

For your lifelong learning solutions, visit **www.cengage.co.uk**

Purchase your next print book, e-book or e-chapter at **www.cengagebrain.com**

Printed in the United Kingdom by Ashford Colour Press
Print Number: 08   Print Year: 2023

Throughout my working life I have met many people and I would like to thank every one of them for their help, support and friendliness. I would also like to dedicate my chapters to my wife Michelle and my children Jessica and Adam – sorry for being grumpy when I was busy.

David Hobson

I would like to thank my wife Sharon, daughters Jessica and Kirsten, also my late stepson Peter and stepson Simon, for their patience and support whilst I have worked on this book.

Patrick Hamilton

I would like to thank my family and friends for their support, inspiration and motivation to help me through the many hours of research and writing whilst working on this book.

Graham Baker

CONTENTS

PART 1 Introduction 1

SECTION 1
Health and safety practices in vehicle maintenance 2

1 Introduction 3
2 Regulations 3
3 Emergency first aid 8
4 Where there is a blame there is a claim 9
5 Road testing vehicles 9
6 Health and safety terms and conditions 9
7 Hazards 10
8 Risk assessments 11
9 Work related stress 12
10 Putting customers at risk with part worn tyres 13
11 High voltage vehicles 14
12 Common rail diesel health and safety 14
13 Multiple choice questions 15

SECTION 2
Good housekeeping 16

1 Why good housekeeping is important 17
2 Effects of poor housekeeping in the automotive environment 18
3 Cleaning as you go 20
4 Multiple choice questions 21

PART 2 The motor industry 22

SECTION 1
The motor industry 23

1 Workplace structures and job roles 24
2 Communicating with customers 27
3 Time keeping 30
4 Employment rights and responsibilities 31
5 Personal learning and thinking skills 34
6 Making learning possible through demonstration 36
7 Multiple choice questions 38

SECTION 2
Customer service 39

1 Customer service 40
2 Right first time 40
3 Customer protection 42
4 Multiple choice questions 43

SECTION 3
Skills in materials, fabrication, tools and measuring devices 44

1 Tools and equipment 45
2 Fabrication, tools and equipment 52
3 Multiple choice questions 56

PART 3 Engines 57

SECTION 1
Diagnosis and rectification of engine mechanical related faults 59

1 Introduction to engine terms and conditions 60
2 Diagnosis of engine related faults 61
3 Lack of performance, misfiring and poor running 64
4 Multiple choice questions 73

SECTION 2
Diagnosis and rectification of engine management and emission related faults 74

1 Engine management systems 75
2 Air flow sensor (sometimes called a mass air flow meter) 75
3 Knock sensors 76
4 Coolant temperature sensor 77
5 MAP sensor 78
6 Camshaft and crankshaft sensors 79
7 Exhaust gas recirculation valves 80
8 Oxygen sensors (Lambda sensors, O_2 sensor) 81
9 Catalytic converters 83
10 NO_x sensor and catalyst 84
11 Diesel particulate filters (DPF) 85
12 Variable valve timing (VVT) 86
13 Turbo chargers 88

14 Stop start technology 90
15 Direct petrol fuel injection systems 90
16 Multiple choice questions 92

SECTION 3
Diagnosis and rectification of lubrication and cooling system related faults 93

1 Lubrication systems 94
2 Cooling systems 97
3 Multiple choice questions 99

SECTION 4
Overhauling engine units 100

1 Purpose and function requirements of engines 101
2 Timing belts and chains 102
3 Camshafts 103
4 Cylinder head removal and valve inspection 104
5 Valve stem oil seals 106
6 Replacing hydraulic lifters 107
7 Piston protrusion 107
8 Bent connecting rods 108
9 Piston rings 108
10 Cylinder bore inspection 109
11 Cylinder bore finish 109
12 Crankshaft end float 109
13 Crankshaft removal and inspection 110
14 Plastigauge 110
15 Multiple choice questions 112

SECTION 5
Hybrids and alternative fuels 113

1 Hybrid vehicles 114
2 Extended range/plug in hybrids 118
3 Electric vehicles 119
4 Hydrogen vehicles 121
5 Fuel cell vehicles 121
6 Multiple choice questions 122

PART 4 Chassis systems 123

SECTION 1
Diagnosis and rectification of steering related faults 125

1 Introduction 126
2 Airbags 126
3 Power assisted steering (PAS) systems 128
4 Four wheel steering 136
5 Steering geometry 137
6 Multiple choice questions 143

SECTION 2
Diagnosis and rectification of suspension related faults 144

1 Suspension terms 145
2 The purpose of suspension 145
3 Passive and adaptive suspension 146
4 Electronic struts/dampers 148
5 Air suspension systems 149
6 Diagnosis of airbag suspension systems 152
7 Magneto-rheological dampers 152
8 Hydro-pneumatic suspension 154
9 Hydro-pneumatic suspension and braking systems 155
10 Multiple choice questions 157

SECTION 3
Diagnosis and rectification of braking related faults 158

1 Diagnosis and rectification of braking related faults 159
2 Brake fluid 160
3 Pads, discs and calipers 161
4 Roller brake tester 165
5 Calculating braking efficiencies 167
6 Electronically controlled handbrake systems 169

7 Anti-lock braking systems 170
8 Vehicle stability control (VSC) 175
9 Regenerative braking 176
10 Anti-lock brakes and associated systems diagnosis 177
11 Multiple choice questions 179

SECTION 4
Overhauling steering, braking and suspension units 180

1 Overhauling steering systems 181
2 Overhauling suspension systems 184
3 Overhauling braking systems 185
4 Multiple choice questions 189

PART 5 Transmission systems 190

SECTION 1
Clutches 192

1 Clutches 193
2 Clutch faults and diagnosis 194
3 Multiple choice questions 196

SECTION 2
Manual gearboxes 197

1 Introduction 198
2 Manual gearbox: faults 199
3 Multiple choice questions 200

SECTION 3
Four-wheel drive 201

1 Four-wheel drive (4WD or 4 × 4) 202
2 Two-speed transfer box 202

3 Differential locks 203
4 Viscous coupling (VC) 204
5 Multiple choice questions 207

SECTION 4
Automatic gearboxes 208

1 Automatic transmission 209
2 Fluid flywheel 209
3 Torque converter 210
4 Honda six-speed automatic manual transmission (AMT) 213
5 Direct shift gearbox (DSG) 215
6 Epicyclic (planetary) gear trains 216
7 Automatic gearbox mechanical system 218
8 Automatic gearbox hydraulic system 219
9 Oil cooler 221
10 Gear selection (automatic transmission) 221
11 Electronic control 222
12 Continuously variable automatic transmission (CVT) 224
13 Control 226
14 Semi-automatic transmission 227
15 Electronic transmission control 229
16 Testing and fault diagnosis 230
17 Diagnostics: automatic transmission: symptoms, faults and causes 231
18 Multiple choice questions 232

SECTION 5
Drive line 233

1 Drive line shafts and hubs 234
2 Differential 234
3 Limited-slip differential (LSD) 235
4 Inter axle (third) differential 237
5 Drive line, shafts and hubs: symptoms, faults and causes 237
6 Removing and replacing a cv joint boot 239

7 Final drive and differential diagnostics: symptoms, faults and causes 241
8 Multiple choice questions 241

SECTION 6
Electronic components 242

1 Sensors and actuators 243
2 Multiple choice questions 247

PART 6 Electrical systems 248

SECTION 1
Electronic principles, lighting and infotainment systems 250

1 Electronic principles 251
2 Electronic systems – abbreviations andsymbols 251
3 Waveforms 252
4 Resistance 253
5 Capacitance 254
6 Switches and relays 255
7 Sensors 257
8 Transistors 258
9 Multiplex 260
10 Lighting circuits and regulations 262
11 Diagnostics: lighting – symptoms, faults and causes 266
12 Infotainment system 267
13 Multiple choice questions 272

SECTION 2
Starting and charging systems 273

1 Types of starter motor 274
2 Permanent magnet starters 275

3 Testing/diagnosing starter motor faults 276
4 Charging systems 278
5 Testing/diagnosing alternator faults 279
6 Liquid cooled alternator 281
7 Smart charging systems 282
8 Stop/start technology 282
9 Multiple choice questions 283

SECTION 3
Vehicle body electrical systems 284

1 Auxiliary electrical systems 285
2 Windscreen wiper systems 285
3 Electrically operated windows andmirrors 288
4 Screen heating 291
5 Power seats 292
6 Central locking systems 293
7 Heating, cooling and airconditioning 294
8 Multiple choice questions 303

PART 7 Body repair 304

SECTION 1
Mechanical, electrical and trim components 305

1 Body repair health and safety 306
2 Body repair 307
3 Vehicle manufacture 308
4 Final adjustments to body repair 312
5 Common electrical components 313
6 Adding modifications to vehicles 319
7 Removing and replacing trim components 321
8 Review 328
9 Multiple choice questions 329

FOREWORD

Welcome to this very latest edition in the Vehicle Maintenance and Repair series, popularly known as the 'Brooks Books'. For more than 40 years these books have helped many tens of thousands of motor vehicle students to gain and conveniently record knowledge in the exciting world of automobile engineering.

Although the basic essentials, such as 'suck, squeeze, bang, blow' must remain the same, automotive technology and the way in which it is taught is constantly evolving. Similarly so with the books which from the very beginning have been frequently subject to technical revision and variation of content to suit ever changing needs.

This latest full colour edition with a high proportion of specially drawn illustrations, maintains the original ease of use and understanding, but adds considerable value by including helpful practical tips and guidance to useful websites – all designed to stimulate interest and remain a valuable source of reference. The team of authors, all practicing and experienced lecturers in motor vehicle work, clearly show that they understand and accurately cater for the needs of both students and teachers.

Roy Brooks, Editor

ACKNOWLEDGEMENTS & IMAGE CREDITS

The publisher wishes to thank the following companies:

Benfield – Honda in Stockton on Tees

Stratstone – Land rover in Stockton on Tees

Sytner – Mercedes Benz Teesside

Teesside Motor Factors

T.A. Flavell and Sons LTD

Image credits

ABOUT THE AUTHORS

David Hobson

David left school at 16 and started an apprenticeship in his local garage – Bluebell garage. He worked for the same company for 17 years, and he then started a new career teaching students the subject of motor vehicles at City Centre Training. After 2 years David then secured a job at Darlington College where he has taught for 5 years. He has taught a range of qualifications from entry level to level 3. David says he enjoys passing on his skills and knowledge to students in order to assist them secure an apprenticeship, after all he was once in their shoes.

Patrick Hamilton CertEd MA(Ed) LCGI AAE MIMI EngTech MIMechE QTLS FIfL

Patrick has over 20 years light and heavy vehicle practical experience gained at leading vehicle manufacturers' main dealerships, including 6 years in the Royal Air Mechanical Transport Servicing Section. He is currently Head of the School of Engineering at West Suffolk College and teaches on automotive and engineering courses. Patrick has experience of teaching on a range of light vehicle and heavy vehicle programmes ranging from entry level to Level 4. He has also worked for the Sector Skills Council writing some of their QCF units and has been a technical consultant and author for City and Guilds and the IMI Awards, helping to develop and write their qualifications.

Graham Baker BA, IEng, AAE, MIMI, MSOE, MIRTE

Graham has over 23 years of practical experience working on light/heavy vehicles and tracked vehicles in the Royal Electrical Mechanical Engineers (REME). In 2006 he became a lecturer at York College teaching automotive and engineering course to students from entry level to Level 3. In 2008 he became the Team Leader of the Motor Vehicle Department at York College and the IMI Awards Centre Coordinator. Graham is now working at IMI Awards as a Qualifications Developer.

QUALIFICATION MAPPING GRID

The following grid shows some of the units from awarding bodies which this book either partially, or completely, map across to. It is designed to aid those delivering the qualifications to be able to see at a glance.

The Foundation Learning units have no generic QCF numbers, unlike the level 2 and 3 qualifications, that form a part of the apprenticeship frameworks. For this reason either the awarding body unit number, or in the case of ABC Awards, their own QCF unit number has been used.

	IMIAL	City & Guilds
Health, safety and good housekeeping in the automotive environment	G0102	001 051
Maintain Working Relationships in the Motor Vehicle Environment	G3	003 053
Tools, equipment and materials	G4	004 054
Enable Learning through Demonstration and Instruction	G6	006 056
Identify and Agree the Motor Vehicle Customer Needs	G8	008 058
Inspect Light Vehicles using Prescribed Inspection Methods	LV0506	105 155
Diagnose and Rectify Light Vehicles Engine and Component Faults	LV07 LV11.1	107 157 111 161
Diagnose and Rectify Light Vehicle Chassis System Faults	LV13, LV11.3	108 158 131 181
Diagnose and Rectify Light Vehicle Transmission and Driveline System Faults	LV13, LV11.2	163 121 171 113
Diagnosing and Rectifying Vehicle Auxiliary Electrical Faults	AE06	406 456
Removing and Fitting Basic Light Vehicle Mechanical, Electrical and Trim (MET) Components and Non-permanently Fixed Vehicle Body Panels	BP18	218 268

Students may also need to complete Personal Learning and Thinking Skills (PLTS) and Employment Rights and Responsibilities (ERR) – The awarding body should provide details as to how this is completed and assessed.

Note: Aspects of optional units are contained within relative sections of the workbook. It should be remembered that whilst the main learning objectives of the qualification have been included, it is important to refer to the latest QCF structure and units to ensure full coverage.

ABOUT THE BOOK

 You may want to consider a first aid course. You are likely to find courses running in your local area.

Activity boxes provide additional tasks for you to try out.

 http://www.hse.gov.uk/riddor

Web link boxes suggest websites to further research and understanding of a topic.

 When trying to diagnose an intermittent fault, a good place to start is to question the customer to gather more evidence.

Tip boxes share author's experience in the automotive industry, with helpful suggestions for how you can improve your skills.

 If you are unsure about using and storing any products, always ask your supervisor.

Health and safety tip boxes draw your attention to important health and safety information.

 Often a 'wiggle test' can help find intermittent faults – wiggling possible poor connections and wires.

Diagnostic boxes provide advice and guidance when carrying out diagnostic procedures on a vehicle.

Online lecturer resource Check answers to all of the student activity questions in this book. Please register here for free access: http://login.cengage.com

Multiple choice questions

Choose the correct answer from a), b) or c) and place a tick [✓] after your answer.

1 **How should a part worn tyre be marked?**

   a) Permanently with an ink stamp [ ]
   b) Permanently by cutting into the sidewall [ ]
   c) Permanently by branding the sidewall [ ]

2 **When was the Health and Safety at Work Act introduced?**

   a) 1984 [ ]
   b) 1974 [ ]
   c) 1999 [ ]

Multiple choice questions are provided at the end of each chapter. You can use questions to test your learning and prepare for assessments

Learning objectives

After studying this section you should be able to:

- **Complete a risk assessment.**
- **Understand the procedure for reporting an accident.**
- **Identify risks that may exist for your particular job role.**

Learning objectives at the start of each chapter explain key skills and knowledge you need to understand by the end of the chapter.

A QUICK REFERENCE GUIDE TO THE QUALIFICATION

Generally, students will attend college full time to achieve a Vocationally Related Qualification (VRQ) or part time as part of an apprenticeship, gaining a Vocational Competence Qualification (VCQ) and a VRQ as well as the required functional skills. Motor vehicle qualifications have changed in recent years, allowing a more flexible approach to the selection of units and their delivery. This has been developed by the Sector Skills Council (SSC) for the industry area concerned. The SSC for the automotive industry is part of the IMI group. The SSC is employer-led and acts in response to employer needs. It does this by producing National Occupational Standards (NOS), developed by employer partnerships and working parties. The NOS describe the different functions carried out by people working throughout the range of sectors in the industry. The Skills, Knowledge and Competency requirements are identified for all the levels of technical, parts, sales and operations management.

Qualification Credit Framework (QCF) units have been developed from the NOS. These are the common units which awarding bodies, such as IMI Awards, City and Guilds and Edexcel Btec use to develop their qualifications. This creates a method of standardisation across all awarding bodies.

One of the advantages of the QCF units is that a learner can complete individual units at one training provider using a specific awarding body. They may, for whatever reason, have to move to another training provider who uses a different awarding body, whereby the accredited units are transferable.

The QCF units cover Competency, Skills and Knowledge. Each unit is allocated a 'Credit' value. A predetermined number of 'Credits' are required to achieve a qualification. The QCF units and

'Credits' are transferable across awarding bodies, allowing the learner to build and complete their VRQ and/or Vocational Competence Qualification (VCQ) even if they move around the country.

The Knowledge units cover the technical understanding of the subject area. The Skills units show that the learner is able to carry out practical tasks to a required standard. These units are designed to be delivered in a college and training provider environment. The Competency units show that a learner not only has the required skills but that they are now able to diagnose faults and perform tasks independently within given timescales and with limited support and guidance. The Competency units can only be completed in the workplace.

The VRQ is specifically designed for a training environment-based delivery on either a full or part time basis. With the use of the QCF units it is divided into two areas: those of Skills and Knowledge. A variety of qualifications have been designed to meet the VRQ criteria which are constructed with a mixture of the QCF units. These can provide learners with practical skills and knowledge preparing them for work in the automotive industry.

The Vocational Competence Qualification (VCQ) (originally known as the NVQ) covers the Competency requirements needed for an individual to satisfactorily complete tasks in the workplace which meet the unit requirements.

Employers and learners can select QCF units which meet their business and strategic needs.

This workbook has been designed to assist the learner in developing their knowledge and skills and to provide support for final competence, in Light Vehicle Maintenance and Repair at Level 3.

PART 1 INTRODUCTION

USE THIS SPACE FOR LEARNER NOTES

SECTION 1
Health and safety practices in vehicle maintenance 2

1 Introduction 3
2 Regulations 3
3 Emergency first aid 8
4 Where there is a blame there is a claim 9
5 Road testing vehicles 9
6 Health and safety terms and conditions 9
7 Hazards 10
8 Risk assessments 11
9 Work related stress 12
10 Putting customers at risk with part worn tyres 13
11 High voltage vehicles 14
12 Common rail diesel health and safety 14
13 Multiple choice questions 15

SECTION 2
Good housekeeping 16

1 Why good housekeeping is important 17
2 Effects of poor housekeeping in the automotive environment 18
3 Cleaning as you go 20
4 Multiple choice questions 21

SECTION 1

Health and safety practices in vehicle maintenance

USE THIS SPACE FOR LEARNER NOTES

_____

_____

_____

_____

Learning objectives

After studying this section you should be able to:

- **Complete a risk assessment.**
- **Understand the procedure for reporting an accident.**
- **Identify risks that may exist for your particular job role.**
- **Understand vehicle systems can have serious dangers.**

Key terms

Hazard   Something that has the potential to cause harm or injury.
Risk   The possibility of suffering harm, danger or loss.

www.direct.gov.uk/sickpay

http://www.hse.gov.uk/legislation/hswa.htm

http://www.hse.gov.uk/riddor

http://www.hse.gov.uk/risk

http://www.tyresafe.org

INTRODUCTION

Health and safety in your workplace

As you are studying a Level 3 qualification you should already be aware of health and safety, good housekeeping, the benefits of keeping a workshop clean and the dangers involved in having an unsafe working environment. You should also be aware of dealing with spillages, keeping tools and equipment in good condition and working safely at all times. If you would like to recap what you have already learnt about health and safety there is a lot of information and learning tasks for you to try in *Vehicle Maintenance and Repair Level 1* and *Light Vehicle Maintenance and Repair Level 2* published by Cengage Learning as part of the Vehicle Maintenance series.

Working in the motor industry can be dangerous if you do not learn to work safely and use your common sense. It is important to be careful and think through what you are doing to reduce the chance of having an accident.

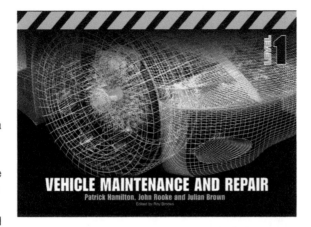

VEHICLE MAINTENANCE AND REPAIR
Patrick Hamilton, John Rooke and Julian Brown
Edited by Roy Brooks

LIGHT VEHICLE MAINTENANCE AND REPAIR
Patrick Hamilton, John Rooke and Robert Sharman
Edited by Roy Brooks

By working safely and sticking to best practices from the start of your career, you can avoid the possibility of suffering long term illnesses and conditions later in your job and in your old age. You need to be aware that conditions like repetitive strain injury can develop slowly and can affect you later in life.

As a simple guide you should be taking care to lift correctly, use correct personal protective equipment (PPE) including dust masks and ear and eye protection when needed and use lifting equipment or ask for help rather than just struggle on. You must always take care of your own health, hygiene and safety. Here we look at health and safety matters that affect you at work and explain some of the regulations. These exist to make sure that you work in good, safe conditions. All major regulations have developed from the legislation in the Health and Safety at Work Act 1974.

 http://www.hse.gov.uk/legislation/hswa.htm

REGULATIONS

As you train you will need to learn and understand the safety regulations that apply to your job. You and your employer are both responsible for following workshop safety regulations.

 TIP You will find copies of the regulations displayed in the workshop. For your own safety, read them.

Workshop regulation posters

Under the Health and Safety at Work Regulations, you have certain duties. Investigate these duties and list some below.

- _____
- _____
  _____
- _____
- _____

If you do not follow the Health and Safety at Work Regulations, you could be taken to court. For example, if you used a bench grinder without wearing the goggles provided, you would be breaking the law, because you would not be taking reasonable care of your own health and safety.

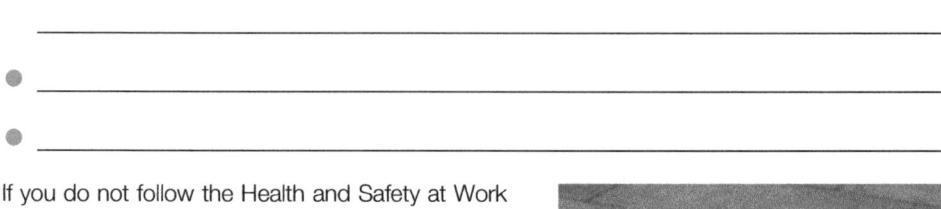

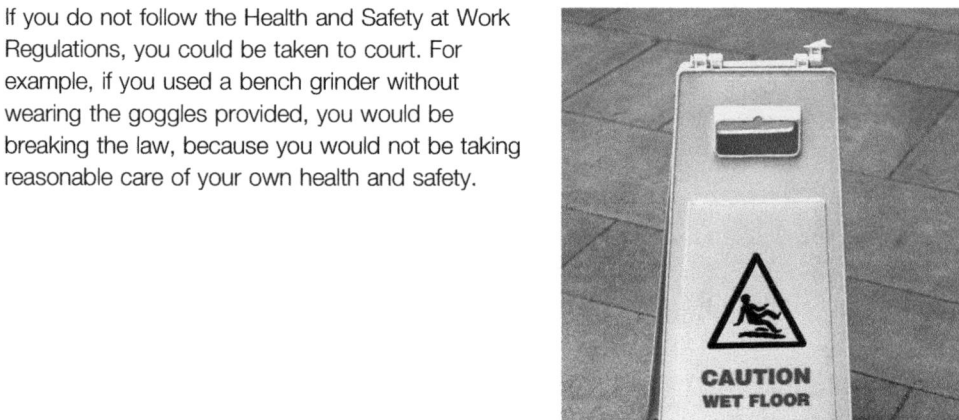

Work with your employer and follow safe systems of work

TIP NEVER tamper with items of PPE or remove safety signs. Someone could have an accident as a result of your actions.

Control of Substances Hazardous to Health (COSHH)

If you are working in a dangerous environment then you may need to complete a COSHH assessment in order to consider control measures.

Complete the table below for a COSHH assessment.

Product	Risk/Hazard	Possible control measures
Petrol, diesel, oils, greases and paraffin	_____ _____	_____ _____ _____
Brake cleaner	Contact with skin and eyes could cause irritation	_____ _____ _____
Oxy acetylene bottles	_____ _____	_____ _____ _____
Detergents or floor cleaners	Contact with skin and eyes could cause irritation	_____ _____

You are working in a small garage and your supervisor asks you to complete a task using a solvent-based product. What are the dangers of completing this task and what could you do about it?

_____
_____
_____
_____

Take note of COSHH labels on products that you may use

If you are unsure about using and storing any products, always ask your supervisor.

 WWW http://www.hse.gov.uk/coshh

Reporting of Injuries, Diseases and Dangerous Occurrences Regulations 1995 (RIDDOR)

All employers owe their employees a duty of care and should therefore provide a safe working environment along with safe tools and equipment. They should also provide PPE and have control measures to reduce the risk of injury as far as reasonably practicable. Employers should provide correct training to prevent employees being injured.

Your employer may have to pay you sick pay if you are entitled to it. Where can you find details of reporting sickness and sick pay?

_____

What situations must your employer or supervisor report to RIDDOR and keep records of?

- _____
- _____
- _____
- _____

_____

An employer must report injuries to RIDDOR which may include amputations, blindness, broken limbs or ribs, unconsciousness, resuscitation and if someone was admitted to hospital for more than 24 hours.

Give an example of when may you need to report a disease to RIDDOR?

WWW http://www.hse.gov.uk/riddor

The employer should also report any injury that prevents an employee completing their work for more than seven days. (On 6 April 2012 this changed from three days to seven days, although a record should still be kept of over three-day accidents.)

What is the purpose of an accident book? Why is an accident book helpful for an employer?

_____

_____

_____

_____

_____

_____

Locate an accident book in your workplace or training centre and describe the details that need to be added in the event of an accident.

- _____
- _____
- _____
- _____

Environmental Protection Act

The Environmental Protection Act is defined as release of any substance into air, water or land as a result of any process which causes harm to humans. You must be very careful as to how you dispose of the waste you produce in the automotive environment.

Scenario

Someone calls by your garage and offers to take your scrap metal or other waste away. What should you do? Explain your answer.

_____

_____

_____

Bench grinder

Be aware of other tasks being completed in the workplace. It would be unwise to use a grinder if someone else in the workplace was draining a fuel tank.

Abrasive wheel regulations

In order to reduce accidents with abrasive wheels and therefore grinders found in the automotive environment, it is very important that they are kept in good condition and checked at regular intervals. Someone in the workplace should have received training on abrasive wheels so that they can change and mount abrasive wheels on a variety of machines and equipment. You should not attempt to replace any abrasive wheel until you have received adequate training.

You should do some preliminary checks before you use a bench grinder. List three safety precautions that should be carried out.

1 _____

2 _____

3 _____

Vehicle air conditioning regulations

To meet European legislation anyone involved in handling (this includes removing, storing and re-charging) refrigerant in vehicle air conditioning systems must be suitably trained and have a certificate to do so. After July 2010 all MAC (Mobile Air Conditioning) technicians must have achieved, as a minimum requirement, a refrigerant handling qualification which meets current EC Regulation.

Why do you think it is important to be trained and certified to work on air conditioning systems?

_____

_____

Air conditioning machine

_____

_____

If you are regularly involved with removing and replacing air conditioning components then you may want to ask your supervisor to allow you to complete some continuing professional development (CPD) in the way of an air conditioning gas handling course.

Investigate air conditioning gas handling courses run at local colleges. List the main things you will learn on an air conditioning gas handling course.

_____

_____

_____

_____

Noise at work regulations

Loud noise can be constant or sudden and it can cause hearing loss that can be temporary or permanent. There are laws that set limits on how loud and safe noise levels should be before precautions are necessary (Control of Noise at Work Regulations 2005). Deciding if you have a problem at work depends on how long you are exposed to noise and how loud the noise is. As a general rule you may need to take precautions If you work in a noisy environment, use noisy machinery for more than half an hour per day or need to raise your voice when talking to someone less than two metres away. Hearing loss may develop over many years and you may develop tinnitus (ringing, buzzing or humming in the ears), a distressing condition which can lead to disturbed sleep.

Noise is measured in decibels (dB) an A-weighting sometimes written dB(A) is used to measure average sounds, it can also be measured as a C-weighting or dB(C) which is used to measure peak or sudden noise levels.

Mandatory signs should be displayed to show hearing protection should be worn

At what noise level should hearing protection be provided for those who request it?

_____

If noise levels are above 90dB then staff and visitors should be wearing mandatory hearing protection.

Simplified Noise Level Reduction, sometimes referred to as single number rating (SNR), is a rating given to show the amount of hearing protection that ear defenders can provide. It is a rating given to an ear defender that is subtracted from the overall sound measurement to calculate the sound pressure level at the ear when wearing a particular ear defender. Ear defenders should bring noise level down to around 75–80dB(A).

How can ear defenders become dangerous?

_____

_____

Fooling around in the workplace is the biggest cause of accidents and often ends with disciplinary action.

What should you do if a customer wants some personal belongings from a vehicle in the workshop?

_____

_____

What should you do if a customer insists on seeing a fault on their car?

_____

_____

_____

_____

_____

EMERGENCY FIRST AID

You are the first and only person to attend an accident in your workshop. Discuss this situation with your classmates and write some points as to what you should do.

- _____
- _____
  _____
- _____
- _____
  _____
- _____
- _____
  _____
- _____
- _____
  _____

 You may want to consider a first aid course. You are likely to find courses running in your local area.

http://www.sja.org.uk/sja/default.aspx

http://www.bbc.co.uk/health/treatments/first_aid

Protecting yourself when helping others

When helping others it is very important that you protect yourself, not only from perhaps the machinery or electricity that caused the accident, but you need to keep yourself safe when dealing with injuries.

If you have to deal with a victim of an accident what should you consider in order to protect yourself when dealing with their injuries?

_____

_____

How should you protect yourself when dealing with an injured person?

_____

_____

_____

Wear gloves when dealing with accidents involving blood

WHERE THERE IS A BLAME THERE IS A CLAIM

We are now living in a society where a blame culture has led to many people wanting compensation for injuries at work. If you feel you are entitled to compensation you would need to make a claim within three years of the date of the accident, you will probably need the assistance of a claims specialist to help you and your employer should have the correct insurance for such occurrences. Your employer should display the insurance information within the workplace and give you the details if you ask for them.

What is the purpose of compensation being paid after an accident?

_____

_____

 TIP Claiming compensation is not an opportunity to just claim some 'free money'.

ROAD TESTING VEHICLES

You may be expected to road test a vehicle that has just been repaired; this could be part of a quality control procedure in order to check the faults before repair. Ensure you have removed all tools and equipment from under the bonnet; try the brakes on the vehicle before setting off, especially if you have completed a repair on the brakes. Ensure you wear your seat belt and keep your speed low around the garage. It is important that you obey traffic laws, don't drive faster and further than you need to and be careful not to make excessive turns or harsh braking.

You should also check with your supervisor to ensure that you are insured to road test vehicles on the company insurance policy. Although you are driving on company insurance if you have penalty points on your own licence then this may affect whether you are allowed to drive or not.

 TIP If you are caught speeding in a customer vehicle it is you that will receive the fine, or possible points on your licence or even a ban – not the customer.

It is common for garages to arrange collection of vehicles from the customer's home or place of work for their convenience.

 Scenario

Your supervisor has asked you to collect a customer vehicle that is booked into your garage for an air conditioning repair. Quickly checking the vehicle you see that both front tyres are clearly illegal. Think about the conversation you may have with the customer. In the space below, write down what the customer and your supervisor may expect from you and what you would actually do. Discuss your thoughts with your classmates.

HEALTH AND SAFETY TERMS AND CONDITIONS

Explain the health and safety terms and conditions listed below and on page 10:

Asbestosis – _____

_____

Competent Person – _____

_____

HASAWA – _____

COSHH – _____

Negligence – _____
_____

RIDDOR – _____

Vibration white finger – _____

Carcinogen – _____

Asphyxiation – _____
_____
_____

PAT Testing – _____

Hazard – _____

Prohibition – _____

Ergonomics – _____

HAZARDS

Describe a hazard.

_____
_____

⚡ If you see a hazard at work it must be reported immediately.

Recognising hazards at work

The first step in being safe at work is recognising hazards. There are many types of hazard and in the motor industry we can put them into different categories.

Physical hazard

Physical hazards are often the most common and can be the easiest to spot; they include unsafe conditions that can cause harm.

Give some examples of physical hazards:

● _____

● _____

● _____

● _____

Ergonomic hazards

Ergonomics is, essentially, the study of how things interact with the human body. It is most often used to make products that are easier or more comfortable for a human to use.

How can you describe an ergonomic hazard?

_____
_____

Give two examples of ergonomic hazards and the injuries they could cause:

1 _____

2 _____

Chemical hazards

Chemical hazards, as the name suggests, come from dealing with the many products in the automotive environment that can be dangerous, such as oil and fuel, but also include fumes from exhausts or welding and chemicals such as cleaning products.

Examples of hazards

Have a look in your place of training or work and list three hazards for each of the categories physical, ergonomic and chemical hazards.

Physical

- _____

- _____

- _____

Ergonomic

- _____

- _____

- _____

Chemical

- _____

- _____

- _____

RISK ASSESSMENTS

At some time in your career you may be expected to complete risk assessments in order to keep the workplace safe.

What is a risk assessment and why is it important?

_____

_____

_____

You can never eliminate all risks but often with small procedures or cheap and effective measures, risks can be reduced to protect you and your colleagues.

There are many ways to complete a risk assessment and some may fit organisations better than others, below is a straightforward five-step risk assessment.

Step 1: Identify the hazards

Walk around the workplace and look for something that could cause harm; often managers will speak to staff in order to get their views related to machinery and equipment being used.

Step 2: Decide who may be harmed and how

Think about groups of people in different areas, for example, customers walking in and out of the workshop who could slip or trip, or people working in a parts department lifting heavy stock.

Step 3: Evaluate the risks and decide on precautions

When you have spotted hazards you must think about what to do about them; the law requires you to do everything 'reasonably practicable' to protect people from harm. This could simply be placing a sign in a specific place or buying equipment to assist with lifting heavy items.

Step 4: Record your findings and implement them

Implement your finding so, for example, make sure staff are trained and use the lifting equipment that you have provided, record your results to show a proper check was made and that you have dealt with hazards, share them with the staff, and ask people if the new equipment has made the job safer and easier to complete.

Step 5: Review your risk assessment and update if necessary

Every year or so check that the actions you implemented are still working and keeping staff safe, speak to employees and make changes if necessary. You should then record any changes made.

WWW **http://www.hse.gov.uk/risk**

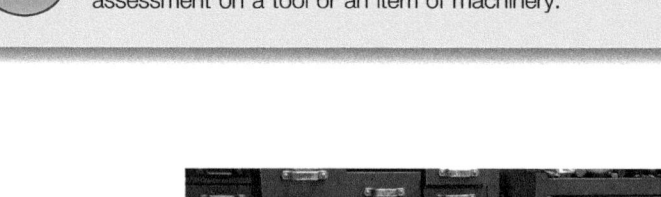
Have a walk around your place of training or work and complete a five-step risk assessment on a tool or an item of machinery.

Untidy tool box

Why is the untidy toolbox above a health and safety risk? Discuss with your classmates.

WORK RELATED STRESS

Whether studying full time or completing an apprenticeship course you could at some time become victim to work related stress. Work related stress could be described as work demands that exceed the individual's capacity and capability to cope. Excessive amounts of stress can result in depression, time off work, reduced productivity and accidents due to human error.

List some items that could cause work related stress in an automotive environment.

- _____
- _____
- _____
- _____
- _____
- _____

Your supervisor may complete a five-step risk assessment related to stress in the workplace;

however, if there doesn't seem to be a policy in place there is one important tool that you must

use if you become stressed. What is this? _____

How would you deal with a difficult task before it causes you stress?

- _____
- _____
- _____
- _____
- _____
- _____

PUTTING CUSTOMERS AT RISK WITH PART WORN TYRES

Vehicle accidents due to the sale of poor quality part worn tyres are becoming more common. It is not illegal to sell part worn tyres but certain conditions must be met in relation to specific markings and the general condition of the tyre.

Why are part worn tyres popular in the UK?

_____

_____

_____

What are the 'hidden' dangers of selling part worn tyres?

_____

_____

_____

_____

How can you identify a part worn tyre?

_____

_____

_____

_____

What is the least amount of tread legally allowed on a part worn tyre that you may sell?

_____

_____

_____

_____

You should also ensure that part worn tyres do not have:

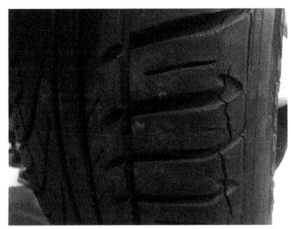
Part worn tyre

● **Any cut over 25mm or 10 per cent of the section width of the tyre (whichever is the greater) on the outside of the tyre, deep enough to reach the ply or cord.**

● **Any internal or external lump, bulge or tear caused by separation or partial failure of its structure.**

● **Any ply or cord exposed internally or externally.**

● **Any penetration damage that has not been repaired.**

It is recommended that a thorough inspection of the tyre be made before fitting to the rim. Tyres that are sold unfitted will need to be inspected with particular care as it is an offence to supply tyres with the defects listed above, even if they only become detectable when the tyre is inflated. Inflation and inspection is recommended.

WWW http://www.tyresafe.org/

Investigate the cost of part worn tyres compared to new tyres – choose five sizes and record your results in the table on page 14.

Tyre size	Price of new tyre	Price of part worn tyre

HIGH VOLTAGE VEHICLES

As a Level 3 student you may become more involved with larger tasks, often completing them on your own. One type of vehicle that you may be asked to work on could be a hybrid vehicle or an electric vehicle.

What are the additional dangers of working on hybrid and electric vehicles?

_____

_____

_____

It is recommended that when you service a hybrid vehicle you should fully remove the key from the ignition. Why is this necessary?

_____

_____

_____

_____

_____

Some manufacturers recommend wearing correctly insulated gloves when working on a high voltage system. Before putting the gloves on what should they be checked for?

_____

_____

_____

COMMON RAIL DIESEL HEALTH AND SAFETY

Today's apprentice or student may be required to work on common rail diesel injection systems. This can have dangers which you may not be aware of. Investigate the main danger of working on a common rail diesel engine.

_____

_____

_____

Never put your hands near to a leak on a high pressure circuit on a common rail diesel engine. The high pressure diesel could cause you serious injury.

Some manufacturers recommend the use of a fuel additive in their common rail injection vehicles to help clean diesel particulate filters and injectors. What health and safety tips would you give to an apprentice adding such an additive to a vehicle?

_____

_____

From time to time a common rail diesel engine may 'regenerate'. This is when the diesel particulate filter is cleaned out during driving. What should you be aware of when a vehicle is 'regenerating'?

_____

When removing a diesel particulate filter, what should you be aware of?

_____

_____

What safety precautions should you take when removing a diesel particulate filter?

_____

 TIP Follow manufacturer's instructions when repairing high pressure systems.

3 **Who is responsible for health and safety in the workplace?**

a) The employer [ ]

b) The employee [ ]

c) All individuals [ ]

4 **What is the main hazard presented by common rail diesel engines?**

a) Excessive fuel pressure [ ]

b) Excessive injector noise [ ]

c) Excessive exhaust fumes [ ]

5 **What is the rating given to the amount of protection that ear defenders can provide?**

a) Sound number rating [ ]

b) Secondary number rating [ ]

c) Single number rating [ ]

Multiple choice questions

Choose the correct answer from a), b) or c) and place a tick [✓] after your answer.

1 **How should a part worn tyre be marked?**

a) Permanently with an ink stamp [ ]

b) Permanently by cutting into the sidewall [ ]

c) Permanently by branding the sidewall [ ]

2 **When was the Health and Safety at Work Act introduced?**

a) 1984 [ ]

b) 1974 [ ]

c) 1999 [ ]

SECTION 2

Good housekeeping

Learning objectives

After studying this section you should be able to:

- Understand that failing to complete basic housekeeping activates can have huge consequences.
- Recognise terms and conditions associated with good housekeeping in the workplace.
- Understand the importance of cleaning as you go.

Key terms

Inflammable   Something that can be ignited or set alight.

Inhalation   Breathing in chemicals and fumes.

Welfare   A term used when discussing well-being of employees.

http://www.popularmechanics.com/cars/how-to/repair/4324365

WHY GOOD HOUSEKEEPING IS IMPORTANT

Many students have the impression that when they start a job or apprenticeship in the motor industry, cleaning the work area is not part of their job and the company just wants someone to clean the workshop without paying high wages. This is certainly not the case: keeping a clean and tidy workplace is very important for health and safety and EVERYONE. Even fully skilled technicians, should be responsible for good housekeeping in the automotive environment.

Under current legislation your employer must provide a safe place for you to work in. This means the equipment you use must be in a good safe condition and the environment you work in should present no obvious hazard to your health. Your employer must also make sure that the workplace is kept as clean as possible. In order for your employer to carry out their duties under the legislation you are expected to help maintain your workplace along with keeping machinery and equipment clean and in good condition – this is referred to as 'good housekeeping'.

Scenario

You have started work in a new job, it is your first day and your supervisor has asked you to tidy the tool store. What would you do in this situation?

Along with good health and safety what are the advantages of having a clean and tidy workplace?

- _____
- _____
- _____
- _____
- _____

Many people are injured due to trips, slips and falls every year. The risk is increased in the automotive environment because the nature of the tasks completed are practical and involve potentially hazardous equipment and products. Give two examples of when spillages can easily occur in the workshop.

1 _____

2 _____

It is important to keep a clean and tidy workplace

Trailing cables, air lines and clutter or blocked walkways are common causes of trips and falls.

What kind of injuries may be sustained from simply tripping over a badly routed cable?

- _____
- _____
- _____
- _____
- _____

Slips, trips and falls are common causes of accidents caused by poor housekeeping in the workplace.

List some of the many preventable ways to prevent slips, trips and falls in relation to good housekeeping.

- _____
- _____
- _____
- _____
- _____

Fire prevention

How much fuel should a motor vehicle workshop store and how should it be kept?

_____

_____

_____

Keep evacuation routes clear and never block fire exits. You should never attempt to clean equipment using petrol, diesel, paraffin, thinners or methylated spirits. Apart from the fact they can cause damage to some materials they are all highly flammable and produce explosive vapours. They can also cause skin and eye irritation and prolonged use could cause dermatitis.

A clean workshop is a safer workshop

Have a walk around the workshop in your place of training or work and look for hazards that should be tidied or cleaned.

EFFECTS OF POOR HOUSEKEEPING IN THE AUTOMOTIVE ENVIRONMENT

Scenario – Oil change

You are asked to complete an oil change on a customer vehicle. Wanting to complete the task in good time you accidently spill some oil but continue to complete the task thinking you can clean up later. You go to the parts department but slip on the oil damaging your knee cap.

Describe how this may affect you, your family, your employer and your workmates using the space provided on the following page.

You

_____

_____

_____

_____

Your family

_____

_____

_____

_____

Your employer

_____

_____

_____

Your workmates

_____

_____

_____

_____

Something as simple as not cleaning up a spillage can have serious consequences for others in the workplace, as well as yourself.

Personal hygiene

Employers must provide a number of welfare facilities to ensure the safety, comfort and welfare of the staff employed. List the welfare arrangements in your place of training or employment.

- _____
- _____
- _____
- _____
- _____

These areas should be kept clean, well ventilated and well lit, students and apprentices should be encouraged to use these areas and clean up after themselves, using rest areas and eating areas can give you the necessary breaks from work to reduce stress and prevent possible contamination of your food and drink. Housekeeping issues within these areas should be reported to your supervisor.

TIP You should ensure you wear correct PPE at all times, it is your responsibility to do this and should include barrier cream or gloves and dust masks as required.

Chemicals can enter your body through 'absorption'. Suggest two ways that this could happen.

1 _____

2 _____

Breathing in chemicals, fumes and dust means they can stay in the lungs and cause a variety of problems, particularly after years of exposure. This is known as _____.

It is possible in the automotive industry to swallow chemicals, usually from contaminated food and drink; this is why it's important not to take food and drink into the workshop and to ensure you wash before taking your breaks. This is known as _____.

A student or apprentice should be aware of the terms listed on page 20 that can be used to describe dangerous substances. Research and explain the terms and draw the safety sign to match the description.

Flammable – _____
_____
_____
_____

Inflammable – _____
_____
_____
_____
_____
_____

Corrosive – _____
_____
_____
_____

Toxic – _____
_____
_____
_____

Reactive – _____
_____
_____
_____

Some substances in the workplace may have a Workplace Exposure Limit (WEL). This sets out the exposure level that must not be exceeded. Information on the WEL for a particular product may be found on a hazard data sheet that should be made available by the product manufacturer.

Not exceeding the WEL does not guarantee health protection as differences in human susceptibility would make this impossible. Read the data sheet and use correct PPE as recommended.

Remember:

- Some products are more harmful than others.
- Some products need large doses before they cause harm.
- Some products harm you quickly.
- Some products take years of exposure before ill effects are seen.
- If you are unsure always ask your supervisor.

CLEANING AS YOU GO

Often when repairing cars, despite using the correct vehicle protection, it is inevitable that you may one day leave dirt on the vehicle. Keeping customers' cars clean will not only show that you have a pride in your work but will also impress your customers and therefore they should return to your garage for future work. Complete the table below to say how you would deal with returning the car to the customer.

Issue with the vehicle	Action required
Finger prints on a window that has been replaced.	_____
Dirty finger marks on a door trim panel.	_____
Spilt brake fluid on a wing.	_____
A fine scratch on a wing edge.	_____
Dirt from your overalls on a seat.	_____
Finger marks on the vehicle paintwork.	_____

Remember, if there are vehicles in the college or workplace that are dusty, never write messages or draw in the dust, this can scratch the paint and cause permanent damage. Advise trainees of this also as it could save costly repairs and disciplinary action.

Disposing of waste in the automotive environment

You may consider that the environment is everything around us, air water and land. Together these provide the conditions to sustain all forms of life including people. We should treat the earth with the respect that it deserves and rather than using the planet as a dustbin we should recycle materials as much as we can, as protecting the environment is important to us all.

What is pollution?

_____

What can cause pollution?

_____

_____

All types of pollution may be considered to be illegal and large fines can result for perpetrators.

Make a list of items found in the automotive workshop which may have a negative impact on the environment:

● _____

● _____

● _____

● _____

● _____

● _____

● _____

● _____

● _____

● _____

You should ensure that these products are disposed of correctly using licensed specialist waste companies and ensure you recycle products wherever possible.

Multiple choice questions

Choose the correct answer from a), b) or c) and place a tick [✓] after your answer.

1 **What is meant by the term inflammable?**

a) The substance can combust [ ]

b) The substance cannot combust [ ]

c) The substance is robust [ ]

2 **What is the term given when chemicals have entered a body possibly due to someone eating contaminated food?**

a) Inhalation [ ]

b) Digestion [ ]

c) Ingestion [ ]

3 **How would you dispose of old tyres?**

a) Household waste [ ]

b) Give to a licensed waste specialist [ ]

c) General waste skip [ ]

4 **What is considered to be pollution?**

a) Contaminating land, rivers or air [ ]

b) Contaminating land, air or sea [ ]

c) Contaminating land, water or air [ ]

5 **What can be a major factor in causing eye strain?**

a) Working in poorly lit conditions [ ]

b) Welding without a mask [ ]

c) Working outside [ ]

PART 2
THE MOTOR INDUSTRY

USE THIS SPACE FOR LEARNER NOTES

_____

_____

_____

_____

_____

_____

SECTION 1
The motor industry 23

1 Workplace structures and job roles 24
2 Communicating with customers 27
3 Time keeping 30
4 Employment rights and responsibilities 31
5 Personal learning and thinking skills 34
6 Making learning possible through demonstration 36
7 Multiple choice questions 38

SECTION 2
Customer service 39

1 Customer service 40
2 Right first time 40
3 Customer protection 42
4 Multiple choice questions 43

SECTION 3
Skills in materials, fabrication, tools and measuring devices 44

1 Tools and equipment 45
2 Fabrication, tools and equipment 52
3 Multiple choice questions 56

SECTION 1

The motor industry

USE THIS SPACE FOR LEARNER NOTES

_____

_____

_____

_____

_____

_____

Learning objectives

After studying this section you should be able to:

- Understand job roles and structures.
- Describe methods of communication with customers.
- Understand personal learning and thinking skills.
- Make learning possible through demonstration.

Key terms

DVSA   Driver and Vehicle Standards Agency.

Discrimination   Being treated unfairly.

Continuing professional development (CPD)   How people maintain their knowledge and skills related to their professional lives.

PLTS   Personal Learning and Thinking Skills. These are essential skills for work and general learning.

WWW   http://www.elyamccleave.com/?p=465

http://www.autocity.org.uk/world-of-work

http://www.acas.org.uk/index.aspx?articleid=1461

http://www.tuc.org.uk/tuc/unions_main.cfm

http://www.nus.org.uk

http://www.theimi.org.uk/careers

WORKPLACE STRUCTURES AND JOB ROLES

Workplace structures could be described as the way that individual departments and managers within an organisation work with each other to meet workplace objectives. Depending on the size of the company, workplace structures can be basic and informal or highly complex.

Independent garage businesses

Small businesses such as independent garages have basic organisational control; this is when perhaps one or two people may take authority of the tasks to be completed – one may be the owner and there may be a supervisor. Most employees are flexible and have skills and knowledge to perform different tasks. Communication between employees and the owner or supervisor is often relaxed and easily co-ordinated. Usually independent garages repair a range of vehicles from different manufacturers.

Dealership businesses

A medium/large sized workplace such as a small dealership may group employees into departments such as parts department, body repair department, car sales and vehicle workshop and service receptionists. Everyone will have a defined role and perform duties according to their area of expertise; they may have to report to departmental managers. This type of structure may be known as a hierarchical workplace where lower levels of hierarchy may take direction from the level directly above.

http://www.theimi.org.uk/careers/careers-resources

In groups compare the different job roles in a dealership and independent garage. Draw diagrams to show the different hierarchy of jobs roles available in each workplace.

An independent garage may have few staff compared to a main dealership

Continuing Professional Development

Continuing professional development (CPD) is basically staff training. This could be a range of training courses from complex vehicle systems, such as folding roofs, to a new piece of equipment or computer application. It would be advisable to complete any CPD that is offered to you as it is something to add to your Curriculum Vitae and can help greatly when repairing and servicing vehicles. If there is a particular part of your job that you feel you need extra training in then you should speak to your supervisor, it may be that the training can be done in the workplace or training centre by people with experience of the topic you need help with.

Make a list below of CPD that you would like to attend.

- _____

- _____

- _____

- _____

- _____

 CHECK MOT means Ministry of Transport
DVSA means Driver and Vehicle Standards Agency
Class of vehicle relates to the type or size of a vehicle.

Popular CPD is training to become a vehicle MOT tester. There are requirements that must be fulfilled in order to become a MOT tester. Firstly you must work in a MOT testing garage. You and your supervisor should then apply to DVSA to start the process. There will be forms to complete and you should match the following criteria:

- **You must have a full driving licence in the classes of vehicles you want to test (e.g. heavy goods or light vehicle).**

- **You must be a skilled technician and have been in full time employment for at least 4 years working on the class of vehicles you want to test.**

- **You should have no 'unspent' convictions for criminal offences connected to vehicle testing or the motor trade or involving acts of violence or intimidation.**

- **You should be of good repute.**

- **For classes 3, 4, 5 and 7 vehicles you must either possess an acceptable vocational educational qualification *or* pass a special multiple choice test set by DVSA which consists of 60 technical questions which must be completed within 1.5 hours.**

 TIP Passing your course can reduce the work needed to become a MOT tester.

- **Once you have passed the DVSA test or passed your course you should attend a 2 or 3 day training course at your local DVSA centre and pass a practical test at your workplace.**

Becoming a MOT tester is a responsible job and you should not take this lightly, every time you test a vehicle you have an obligation to ensure the vehicle is roadworthy and safe, otherwise accidents could occur.

DVSA are strict on MOT policies and may contact you from time to time to check the quality of your MOT tests, you may be asked to attend refresher courses.

 WWW **http://www.theimi.org.uk/courses-and-events**
http://www.dft.gov.uk (and search DVSA)

Future careers

There are many job roles in the automotive industry; the skills and knowledge you will learn as an apprentice may allow you to change career; if you are a full time student you may be able to apply for jobs related to motor vehicle repair but not directly as a vehicle technician. The fact that you have knowledge of vehicle systems will assist you greatly in applying for the many jobs available, often in many years' time.

Complete the table below with a brief description of the job roles stated.

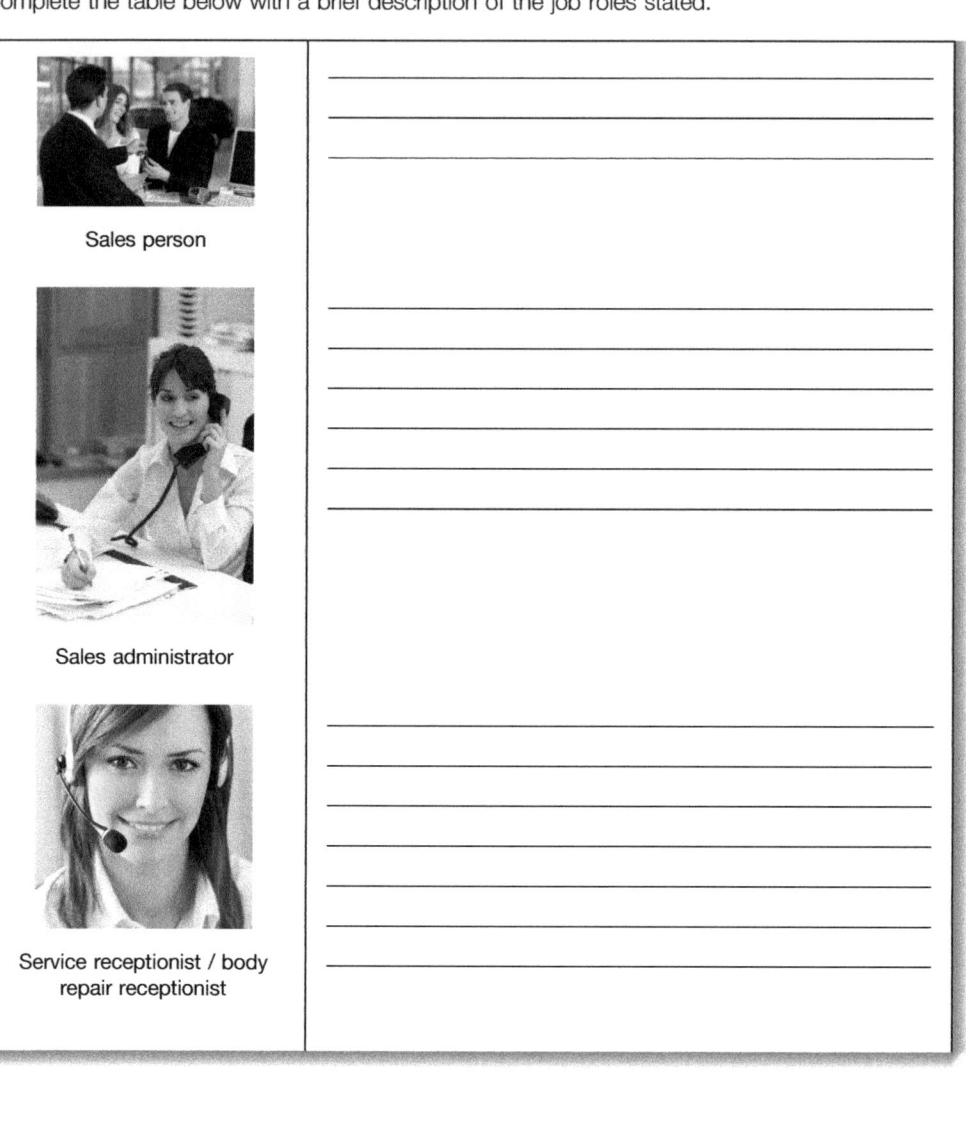

Sales person

Sales administrator

Service receptionist / body repair receptionist

Warranty clerk

Quality controller

Parts department staff

Supervisor

Roadside recovery technician

Car cleaner

RENT A CAR

Vehicle hire clerk

Motor vehicle tutor

COMMUNICATING WITH CUSTOMERS

Methods of communication

Often you will need to deal with customers within a job role in a garage, you should be aware of how to deal with customers in a professional and courteous manner.

Don't interrupt and listen carefully

Be careful of interrupting customers, when they are talking to you about a complaint try to be patient, listen for key facts and wait for an appropriate point to ask further questions. Diagnosing faults correctly involves gathering evidence and a great place to start is by listening to customers. Let them tell you what happened, be interested and give the customer encouragement that you can repair the fault.

Avoid negative questions

Ask questions related to the fault, if the customer doesn't quite understand the vehicle system then don't laugh at their response. If you don't receive the answer you expected ask the question in a different way.

Be aware of differences in technical knowledge

It would be no good telling a customer that their camber angle needs adjusting without explaining what camber angle is. Maybe you could explain with a sketch on a note pad, customers value you being fair with them so they feel they are receiving value for money.

Remember technical problems can cause stress and emotions

Often a customer will become stressed and emotional when their vehicle lets them down, this can make people sound argumentative and angry. It is important to keep calm, never shout back at them and try to be understanding and offer encouraging comments such as 'Don't worry sir, we will soon have this fixed'. You should be able to gauge the situation; there is a difference between a customer that is emotional and one that is abusive. Abusive customers should be asked to calm down or leave.

Keep the customer informed

If you are having difficulty repairing the vehicle or are awaiting parts then keep the customer informed, offer them the chance to make some phone calls and make other arrangements.

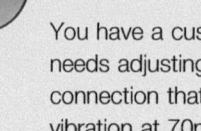

Scenario

You have a customer that has entered your garage, they then claim the 'tracking needs adjusting as the steering is shaking at 70mph'. Firstly you should make a connection that if the tracking is incorrectly set it does not usually cause a vibration at 70mph, and the fault is more likely to be a wheel balance fault. Write down a series of questions that you may ask a customer in order to establish the correct complaint, the checks that you would complete in order to diagnose the fault and the advice you would give to the customer. Share your report with your classmates.

Questions to ask a customer to establish the correct complaint:

- _____
- _____
- _____
- _____
- _____
- _____

The vehicle checks to confirm your suspicions and diagnose the correct fault:

- _____
- _____
- _____

The advice that you would give to the customer:

- _____
- _____

Dealing with angry and abusive customers

From time to time you will face customers that are angry and this can soon turn into an abusive customer. No one should have to face abuse in the workplace – shouting, swearing, threatening and intimidating could all be considered to be abusive behaviour.

Try to resist shouting at any customer, stay calm and ask them to calm down, ask them to stop swearing and listen to what they have to say. Use terms that are to the point such as 'can you stop swearing please' and repeat this as you feel necessary. If the customer is on the phone then say 'can you please stop swearing or I will end the call'. If they continue to swear then end the call, this will allow them a chance to calm down and reflect on their actions. You could say 'I can only help you when you calm down'.

If the customer is emotional and angry then let them speak, empathise with them and perhaps put yourself in their shoes to understand why they are upset, be understanding and thoughtful. If there are other customers present try to remove them from the area, perhaps to an office, and offer them a seat and a drink.

Discuss with your class and write down some ideas on how to deal with irate customers.

- _____
  _____
- _____
- _____
  _____
  _____
  _____
- _____
  _____

WWW http://www.elyamccleave.com/?p=465

On the following images decide if the communication method is appropriate or inappropriate communication.

'Keep talking I am listening'

TIME KEEPING

Many employers run a bonus scheme for motor vehicle technicians and if you work in a main dealership you may be expected to complete warranty work. Most manufacturers ask you to prove the time taken on a warranty job and often complete audits to check. The workplace will have a clock and you should clock on at the start of the job, then clock off at the end. Sometimes this is done electronically on a computer.

Almost every job in the motor trade will have a time allocation attached to it, either set by the employer or the manufacturer. This allows easier quoting of prices to customers to ensure consistency and can also be used to promote a bonus scheme for technicians.

The time is split into 10ths of an hour, so 0.1 equals 6 minutes; for example, if a job has an allocated time of 1.3 hours the actual real time to complete the job is 1 hour and 18 minutes.

Complete the table below to calculate labour time and real time.

Decimal	Real time
0.1	6 minutes
_____	12 minutes
0.3	_____
0.4	_____
0.5	30 minutes
0.6	_____
_____	42 minutes
_____	48 minutes
_____	_____
1.0	_____

If a technician is allocated time of 1.8 hours to complete a job, what is the actual real time for the technician to complete the job?

_____

If a job takes 1 hour 54 minutes to complete, what is the allocated time for the technician to complete the job?

_____

A technician may be contracted to sell 8 hours' worth of labour per day and any extra hours are paid as bonus, this is usually at the same hourly rate as the normal wage. This can be a good way to top up wages; however, it should be noted that completing the task to a high standard should always take preference to making bonus. Rushing tasks can cause mistakes and endanger customers.

 Cutting corners to make bonus is never a good idea.

A technician has completed the work below and is paid a bonus equivalent to the hourly wage when 8 hours of labour have been sold. Calculate the bonus earned for the technician who has an hourly rate of £10.70 per hour.

Task completed	Labour time allowed
A major service	1.5
Timing belt replacement	2.1
MOT Test	1.2
Clutch replacement	5.2
Diagnose alternator fault and replacement	2.0
Total labour time sold	_____
Bonus earned	_____

Sometimes a task may take longer than the allocated time, especially if things go wrong. A snapped bolt or incorrect parts ordered may cause delays, always keep your supervisor and customers informed.

Using a time clock recorder can help you monitor the time spent on each job

 The clock shown above is often used to record time of jobs completed, make sure the start and end times are recorded accurately as it may be audited by the vehicle manufacturer.

EMPLOYMENT RIGHTS AND RESPONSIBILITIES

 Key words:
Rights – Principles or benefits that you are entitled to exercise or claim
Responsibilities – Duties that you are expected to fulfil
Entitlement – Rights that you are allowed or you deserve
Statutory – By law

Many motor vehicle students complete their studies and find a work placement in order to complete an apprenticeship. As an employee you would have rights and responsibilities of which there are many, if you know your rights you will know what you are entitled to and this will prevent you from being treated unfairly. If you know your responsibilities then you and your employer will know what to expect of each other. See if you can list some basic rights and responsibilities below.

- _____
- _____
- _____
  _____
- _____
- _____
- _____

- _____
- _____
- _____
- _____
- _____
- _____
- _____
- _____
- _____
- _____
- _____

Research the following terms and conditions:

Workplace pensions

Holiday entitlement

Bank holidays

Sickpay including statutory sick pay

Working hours

Contract of employment

Breach of contract

Research and describe the terms listed below and on page 33.

Direct discrimination

_____

_____

_____

_____

Indirect discrimination

_____

_____

_____

Discrimination at work

Everyone has the right to be treated fairly whilst at work, it is against the law to treat someone differently for a reason unconnected with the job such as because of their nationality or gender. There are many ways in which you cannot be discriminated against, can you list some?

Harassment

_____
_____
_____
_____
_____

Contract of employment

What is the purpose of having a contract of employment?

_____
_____
_____

If you don't comply and stick to the duties in the contract then this is known as breaking the terms of the contract and could lead to disciplinary action.

Dismissal

If you are dismissed from work it is generally due to wrong doing.

What should you do if you feel you are wrongly dismissed?

_____
_____
_____
_____
_____
_____
_____
_____
_____
_____

Research ACAS – find out who they are, what they do and how they can help in employment disputes.

http://www.acas.org.uk/index.aspx?articleid=1461

Trade unions

Investigate what is meant by the term 'union'.

Describe the meaning of the term trade union.

_____
_____

How can a union membership help you?

_____

_____

There are many unions that you can join and you would be wise to complete some research to see which one would suit you in relation to their policies, priorities and the type of job you undertake. You don't have to join a union and it is entirely down to choice. Listen to advice from colleagues to see if unions are supported in your place of work. If you do join a union you should know who the union representative is and whom to contact should you have issues, this may even include disciplinary issues and often you can take a union representitive to disciplinary meetings with you. There is normally a fee to pay to be part of a union, often taken from your wages each month. By being part of a union you may be able to receive discounts from stores or manufacturers.

You will be surprised how many unions there are to join, including one for students. Ask in your place of training or study about the student union. If this is not recognised in your college/training centre you may be able to become involved in student voice groups, learner councils or become a student representitive.

Research union information and unions connected to the motor industry.

http://www.tuc.org.uk/tuc/unions_main.cfm

http://www.nus.org.uk

PERSONAL LEARNING AND THINKING SKILLS

Personal learning and thinking skills have been developed to help you reflect on the work that you do and hopefully assist you in finding new and easier ways to do things. Study the diagram to the right and think how you may fit into the different catagories.

If you are completing a Level 3 qualification you may need to collect PLTS evidence and log this into a portfolio. The PLTS framework comprises six groups of skills that, together with the Functional Skills of English, Mathematics and ICT, will prepare learners to confidently enter work and adult life as self-assured and capable individuals. The six groups of skills are as follows:

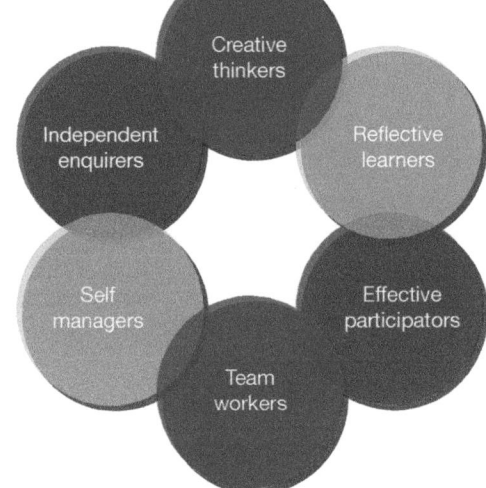

Personal learning and thinking skills

Creative thinker

Young people think creatively and try different ways to tackle problems, often working with others to find imaginitive solutions that are of value. Often a creative thinker will ask questions in order to improve their thinking skills.

Reflective learners

Students may monitor their own performance and invite feedback from others to make changes to training or learning. Students may respond positively to praise, crititism and setbacks, making improvements in behaviour and attitudes when necessary.

Effective participators

Students may play a part in improving situations in learning or working environments, they may identify improvements that will help others as well as themselves.

Young people work confidently with others, often taking responsibility for their own part. Often they will show fairness and consideration to others.

Self managers

Young people organise themselves showing personal responsibility and initiative, often managing risks and demonstrating flexibility when priorities change.

Independent enquirers

This unit shows how young people can plan what to do and how to go about it, they take their own decisions but realise that people have their own beliefs and attitudes.

Complete the information below related to PLTS.

Examples of Personal Learning and Thinking Skills

Creative thinker

Give an example of a problem that you have encountered and how you checked that the problem had been solved.

_____

_____

_____

_____

_____

_____

_____

Reflective learners

Give an example of positive feedback received and an example of a criticism that you have received, describe how you turned the criticism into positive feedback.

_____

_____

_____

_____

_____

_____

Effective participators

Identify improvements that you have suggested that would benefit others as well as yourself.

_____

_____

_____

_____

_____

_____

Teamworkers

Briefly describe a task that you have been involved in where you have contributed to teamwork.

_____

_____

_____

_____

_____

_____

_____

Self managers

Describe a task that you have completed and how you self-managed it, how you organised tools and information, how you managed your time and how you anticipated and managed risks.

_____

_____

_____

_____

_____

Independent enquirers

Give an example of something you have had to research in order to solve a problem and explain the importance of researching information from different sources before making your own decision.

_____

_____

_____

_____

_____

_____

_____

MAKING LEARNING POSSIBLE THROUGH DEMONSTRATION

Practical tasks are learnt quickly through demonstration and when students are involved with the demonstration this can advantage both the demonstrator and the audience. The audience can obviously gain knowledge from the demonstration and the presenter from research related to the task. The presentation is just the start, you should be aware of your audience, health and safety, factors that are likely to prevent learning such as language barriers and not understanding terminology.

What is one of the most important things about completing a demonstration?

_____

_____

_____

Scenario

You have been asked to complete a demonstration to a group of younger students, this could be removal and replacement of a certain part or the use of equipment. Think about how to structure a demonstration or instruction session.

- _____
- _____
- _____
- _____
- _____
- _____
- _____
- _____
- _____
- _____

Demonstration activity

Choose a subject, plan and complete a demonstration to some younger students or your classmates.

The name of the task is ....

The step by step approach to completing the task is:

Step 1 _____

_____

Step 2 _____

_____

Step 3 _____

_____

Step 4 _____

_____

Step 5 _____

_____

Step 6 _____

_____

Step 7 _____

_____

Step 8 _____

_____

Step 9 _____

_____

List the tools you will need to complete the task:

● _____

● _____

● _____

List health and safety requirements and PPE required to complete the task:

● _____

● _____

● _____

Consider the involvement of the audience – what will they be doing?

● _____

● _____

Design a learning game to teach younger students. This could be anything from a computer-based game to a jigsaw or quiz but it must demonstrate a skill to the younger audience.

Multiple choice questions

Choose the correct answer from a), b) or c) and place a tick [✓] after your answer.

1 **What does DVSA mean?**

   a) Driver and Vehicle Operators Service Advisor [ ]

   b) Driver and Vehicle Standards Agency [ ]

   c) Driver Variety of Service Agents [ ]

2 **How many minutes in 0.6 hours?**

   a) 42 [ ]

   b) 27 [ ]

   c) 36 [ ]

3 **What is student feedback?**

   a) Offering advice, positive comments, or criticism to aid future learning [ ]

   b) Offering negative comments and criticism to aid future learning [ ]

   c) Shouting at students when things go wrong in order to get them corrected [ ]

4 **What might you need to become an MOT tester?**

   a) A vocational qualification [ ]

   b) A windscreen checking competence certificate [ ]

   c) Initial assessment results [ ]

5 **What is continuing professional development?**

   a) Continuing training and lifelong learning [ ]

   b) Training until you are a professional [ ]

   c) Working until your training is complete [ ]

SECTION 2

Customer service

USE THIS SPACE FOR LEARNER NOTES

_____

_____

_____

_____

_____

_____

_____

Learning objectives

After studying this section you should be able to:

● Understand the importance of keeping customers informed of progress.

● Recognise that 'right first time' is important.

● Understand customer protection in relation to sale of goods act, warranties and data protection.

Key terms

Intermittent fault   A fault that may be present then not present.
Diagnosis   Determining the fault.
Customer protection   Legal rights of customers.

https://www.gov.uk/data-protection/the-data-protection-act

http://www.oft.gov.uk/business-advice/treating-customers-fairly/sogahome/sogaexplained

CUSTOMER SERVICE

A customer should leave the garage with nothing other than a good impression. List some things that help present a good impression.

- _____
- _____
- _____
- _____
- _____
- _____
- _____

Customers should be informed of progress throughout the job. Occasionally things can go wrong, a part is broken or not delivered by another company, and you must therefore inform the customer in order to let them make other arrangements. Also mark down damage and faults on the job card and ensure customers know about them.

Customer service is paramount to running a successful business, the smallest things can really upset people. For example, a service check light that has been left on after an oil change for many will mean another trip back to the garage to have it cancelled. This may be an easy task for a technician but for some customers it can be difficult and annoying.

One way to be successful is to carry out some advertising. How may a garage advertise its services?

- _____
- _____
- _____

Most advertising costs money but there is one advertising method that can be very effective and is free. What is this and how does it work?

_____
_____
_____

RIGHT FIRST TIME

Right first time is all about diagnosing a fault correctly in order to repair the fault first time. Once you have diagnosed the fault and quoted a price to the customer the last thing you need is to contact them and tell them the fault is not fixed and now you want more money for parts. There is a procedure that should really be followed in order to diagnose faults correctly. Explain what the procedures listed below and on page 41 involve.

Gather further information

_____
_____

Verify the complaint

_____

Repair the fault

_____
_____

Verify the fault is repaired

 TIP Unless the vehicle is under warranty the customer should agree to the cost of repairs before you complete the repair.

A diagnostic tester can be used to read engine fault codes and diagnose engine problems

Using your experience to diagnose a fault

Often a customer will complain of a problem and as you walk through the car park to drive the car into the workshop you should be thinking of obvious faults and a good place to start the diagnosis. It's not always easy to find a fault but complete the table opposite to state good areas to start a diagnosis.

Complaint	Possible fault and a place to start the diagnosis
The vehicle runs ok at speed but cuts out when approaching traffic lights or roundabouts	_____
Both brake lights and the high level brake lights are not working	_____
Brake pedal feels spongy	_____
Heater blower motor only works on top speed	_____
Vehicle is jerking / juddering under hard acceleration	_____
Battery warning light stuck on	_____
Vehicle pulls to the right on braking	_____
A vibration when braking from high speed	_____

Intermittent faults

Describe what is meant by an intermittent fault.

_____

Intermittent faults can be very difficult to locate as you may not be able to duplicate the complaint.

TIP When trying to diagnose an intermittent fault, a good place to start is to question the customer to gather more evidence.

What would you ask the customer?

- _____
- _____
- _____
- _____

CHECK Often a 'wiggle test' can help find intermittent faults – wiggling possible poor connections and wires.

CUSTOMER PROTECTION

Data protection

The purpose of the Data Protection Act is to control the way information is handled and to give customer protection (legal rights) to people who have information stored about them. It is important that data relating to the customers in the workplace is kept secure. What might you consider to be data that should be stored securely?

- _____
- _____
- _____
- _____

WWW https://www.gov.uk/data-protection/the-data-protection-act

Sale of goods act

The sale of goods act gives your customers certain legal rights when they buy goods from you. Goods bought in the automotive trade could be considered to be vehicles or parts. What circumstances would mean a customer could ask for a refund?

- _____
- _____
- _____

TIP Customers have the same rights with second hand parts as they do with new so be careful when fitting parts from a scrap dealer or re-conditioned parts.

WWW http://www.oft.gov.uk/business-advice/treating-customers-fairly/sogahome/sogaexplained

Warranties and guarantees

Many retailers offer a guarantee (or warranty) to customers, if you give a guarantee, you should remember it is legally binding. List the key requirements of a warranty below and on the next page:

- _____
- _____

- _____
- _____

There are laws to protect customers as well as their rights under a guarantee or warranty, so you may be expected to repair the complaint for free. This means that if a customer complains to you about an item or repair that is not fit for purpose, does not match the description, or is not of satisfactory quality, you must deal with their complaint and you cannot force them to use their guarantee if the fault lies with you.

TIP Being friendly with your customers and helping to sort complaints for them can help to keep them as future customers.

Wear and tear

Some items on a vehicle may be subject to 'wear and tear' – this means that they may not be covered under the warranty from the manufacturer and in the case of a fault the customer would need to pay for them.

Make a list of items that would be considered to be 'wear and tear' and probably not covered under warranty.

- _____
- _____
- _____
- _____
- _____

- _____
- _____
- _____

Some used car warranties may have, for example, a £50.00 excess – discuss what is meant by this term.

Multiple choice questions

Choose the correct answer from a), b) or c) and place a tick [✓] after your answer.

1 **What may be considered the best way to advertise your business?**

   a) Leaflets [ ]

   b) Word of mouth [ ]

   c) Internet [ ]

2 **Which Act allows legal rights for customers?**

   a) Sale of Goods Act [ ]

   b) Retail Goods Act [ ]

   c) Service Goods Act [ ]

3 **A customer may have a claim for goods because –**

   a) They have broken them [ ]

   b) They are not fit for purpose [ ]

   c) They cannot afford them anymore [ ]

4 **An item may not be covered under warranty because –**

   a) It's been misused by the customer [ ]

   b) The item failed because the car has not been driven for many months [ ]

   c) The owner lent the car to another driver [ ]

5 **One of the first steps in good diagnosis is?**

   a) Check the vehicle on a 2 post lift [ ]

   b) Gather further information [ ]

   c) Collect the correct tools for the job [ ]

SECTION 3

Skills in materials, fabrication, tools and measuring devices

USE THIS SPACE FOR LEARNER NOTES

_____

_____

_____

Learning objectives

After studying this section you should be able to:

● **Understand how diagnostic tools operate.**
● **Be able to recognise tools and equipment for diagnostics.**
● **Relate to skills in fabrication, tools and measuring devices.**

Key terms

Hygrometer   Used to measure moisture content.
Refractometer   Used to measure coolant, battery electrolyte and screen wash strength.
Oscilloscope   A device used to check for electrical signals/readings.
Fabrication   Building metal structures using cutting, bending, and assembling processes.
Case hardened   Heating steel to harden its surface and improve its strength properties.
LCS   Low carbon steel.
HCS   High carbon steel.

www   http://www.picoauto.com

http://www.wisetool.com/designation/treatment.htm

Describe the use of the tool shown above.

_____

_____

_____

Which other tasks may be completed with this tool?

_____

_____

_____

Some diagnostic tools can be plugged into the vehicle and operated via wireless connection to a laptop or PC. They may also include an oscilloscope feature.

Research different types of scan tools available. Complete the table below to compare uses and costs of different scan tools.

Type of scan tool	Scan tool cost	Scan tool use

Compression tester

The tool shown below is a compression tester, often used to determine the condition of internal engine parts such as pistons, piston rings, cylinder bores and valve condition.

Describe how to perform a dry compression test.

_____

_____

_____

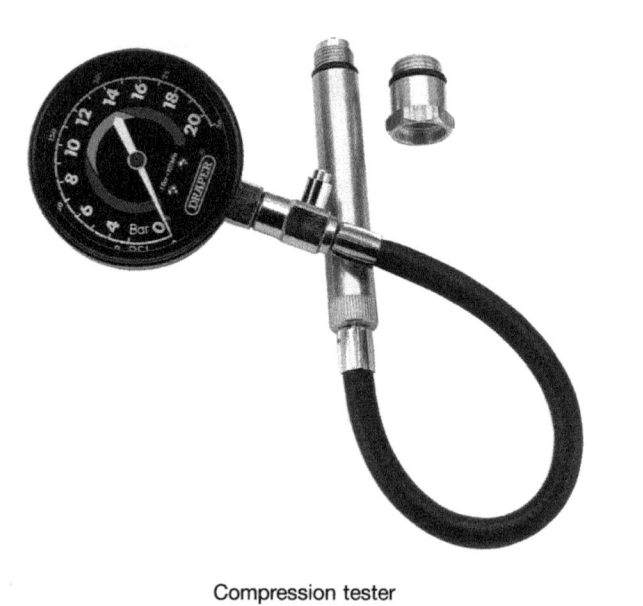

Compression tester

The tool shown below is a cylinder leakage tester, what is the purpose of this tool?

_____

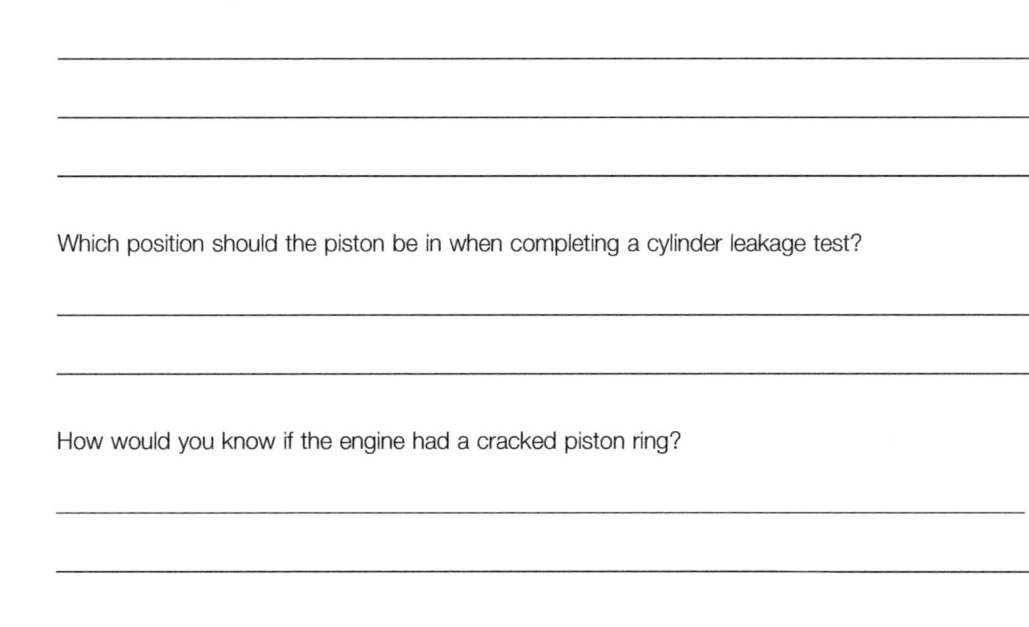

Cylinder leakage tester

Describe how to perform a wet compression test.

_____

_____

_____

How is the tool operated?

_____

_____

_____

The ignition system is disconnected to prevent the vehicle starting but why should the fuel supply be stopped?

_____

_____

_____

Which position should the piston be in when completing a cylinder leakage test?

_____

_____

How would you know if the engine had a cracked piston ring?

_____

_____

How should the fuel supply be stopped?

_____

How would you know if the head gasket had blown?

_____

_____

If a valve was worn, damaged or burnt away where would the air escape to?

_____

_____

Sketch a cylinder bore and piston on the compression stroke; include the valves and cylinder leakage tester, mark on each diagram the air escaping for each fault given.

Worn piston rings   Burnt inlet valve

Damaged Cylinder Head Gasket   Exhaust valve not seating correctly

What is the biggest advantage of using a cylinder leakage tester?

_____

_____

_____

Stethoscope

An old technician's trick that you may know of is to hold a long screwdriver against a part and put it to your ear, a good trick for large noises but not as good as a stethoscope, as shown below. A stethoscope can be a very useful tool to diagnose customer complaints of noise, especially if the noise is difficult to recognise, for example, an internal engine noise, a bearing noise or dashboard rattles.

> Be aware of hygiene when using a stethoscope, ensure the part you put in your ear is clean.

Describe a danger of using this tool.

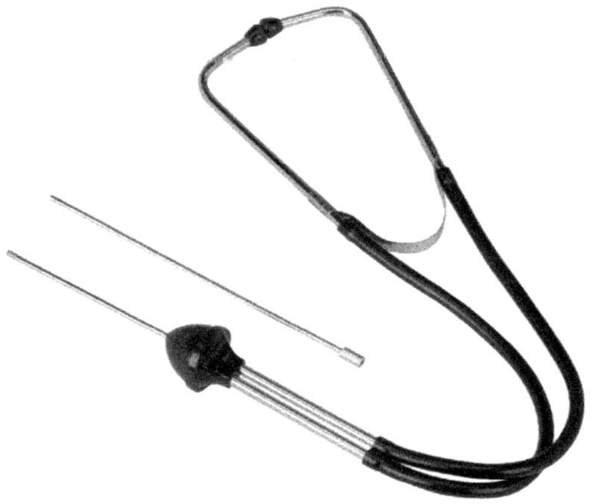

_____

_____

Technician's stethoscope

Multi-meter

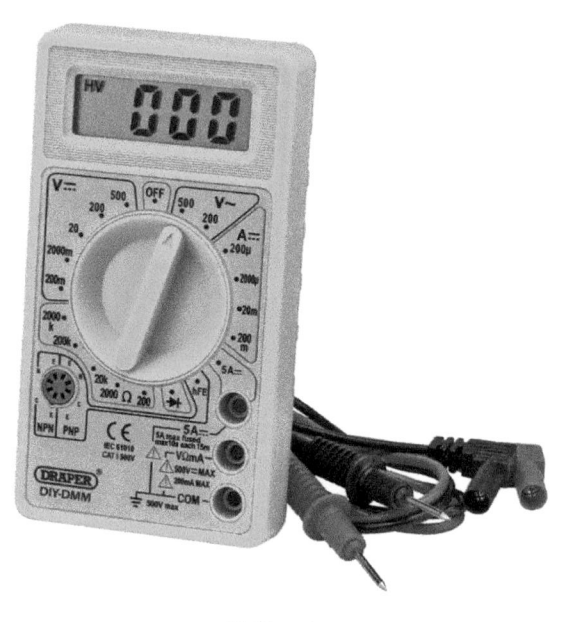

Multi-meter

A multi-meter is a very useful tool for diagnosing electrical faults. Give some examples of tasks that can be diagnosed with this tool.

- _____
- _____
- _____
- _____
- _____
- _____

It is important that you know how to operate the multi-meter correctly and that you connect it without damage.

 **TIP** If you are in doubt about what you are doing then ask your supervisor. By not understanding the readings or not selecting the correct settings you may struggle to diagnose faults correctly.

Draw a diagram to show how a multi-meter can be wired in to measure amps, volts and resistance.

AMPS

VOLTS

RESISTANCE

Test lamp

A test light is an inexpensive and easy-to-use tool to diagnose electrical faults; it can be used to check a circuit for a live feed and earth connection. Add a test lamp to the diagram to check for a live feed and earth connection.

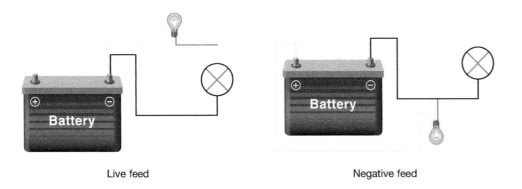

Live feed

Negative feed

Power probe

A power probe is a useful tool for diagnosing electrical faults; it can be used in a similar way to a test lamp, as it can show a negative feed or positive feed. The power probe can also be used to force a live feed or earth into a circuit, You should receive full training on this tool before you use it because incorrect use could result in damage to electrical components which can make the task even more time consuming and expensive.

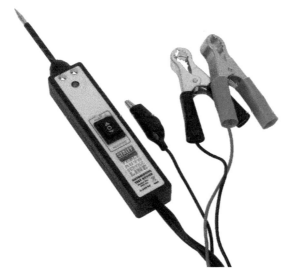

Power probe

Do not use a power probe or a test lamp to check air bag circuits. Accidental detonation can occur.

WWW http://www.expeditionexchange.com/powerprobe

Oscilloscope

What does an oscilloscope do and how can it be used in the diagnosis of vehicle systems?

_____

_____

The oscilloscope is generally connected to sensors that you suspect are faulty; perhaps because the code reader has given you trouble codes. It is useful to diagnose intermittent faults. Often the oscilloscope will be used in conjunction with a laptop computer and the results can be stored, saved and printed.

Mark on the axis of the graph voltage and time as measured by an oscilloscope.

With an oscilloscope you can measure amperage required to crank the engine and also complete a compression test, how would this be done?

_____

_____

_____

_____

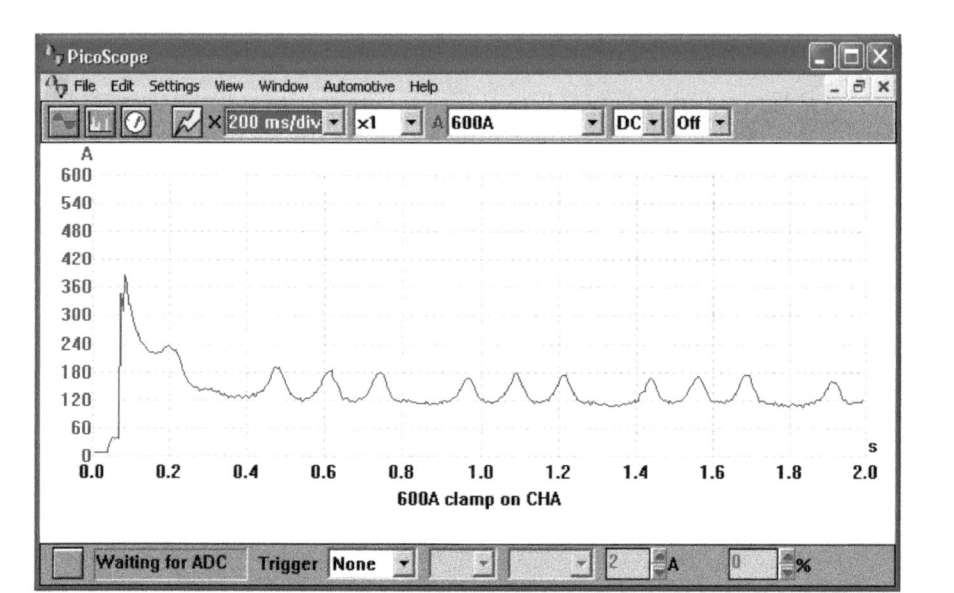

Oscilloscope reading

You can see that after the initial peak cylinder 1 is low on compression.

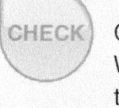

CHECK Oscilloscopes are becoming a popular method of diagnosing intermittent faults. When a sensor may break down only for a split second, this may be too quick for the vehicle engine management system to acknowledge the fault and store a trouble code.

WWW http://www.picoauto.com

Hygrometer

What is meant by the term hygroscopic?

_____

Which product used on motor vehicles is hygroscopic and how does this affect the product?

_____

_____

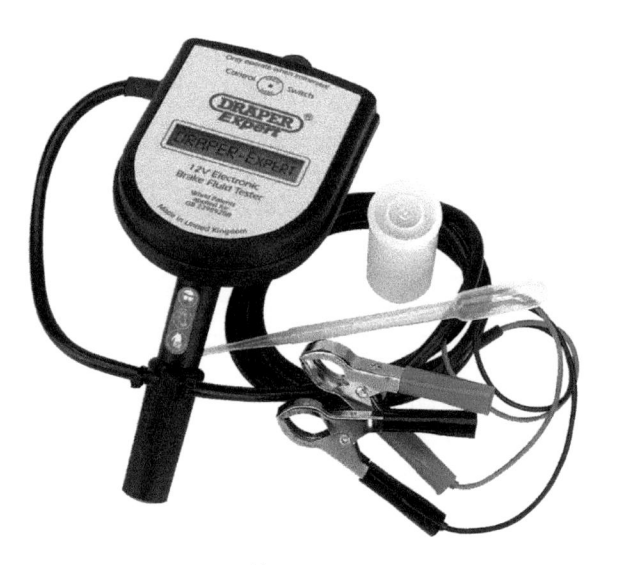

Hygrometer

The hygrometer can measure the moisture content in brake fluid and therefore ascertain when it requires replacement.

Refractometer

A **refractometer** is a precision tool used for measuring anti-freeze strength, battery fluid and screen-wash strength. It can be more accurate than a hydrometer and therefore a good diagnostic tool.

Refractometer in use

How does a refractometer work?

_____

_____

The tool works on the principle that it measures light as it passes through the liquid, as light passes through it will slow down. If that light is passed through at an angle, the substance will 'bend' the light. The angle at which this light is 'bent' can be measured and converted into a refractive index number which can then be converted into a percentage concentration number.

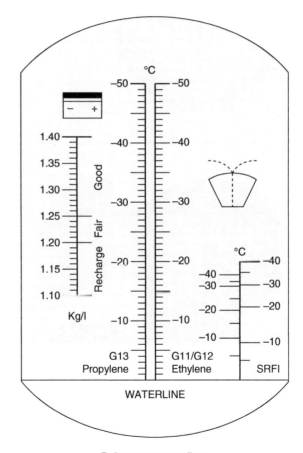

Refractometer readings

FABRICATION, TOOLS AND EQUIPMENT

At some point during a career in the motor industry you will have to fabricate, alter, modify or repair something. This could be anything from making a bracket to drilling out a broken stud. You will have probably studied many of the tools used at Level 1 or 2. As a short recap see if you can name the tools below used in fabrication, measuring and repair.

_____

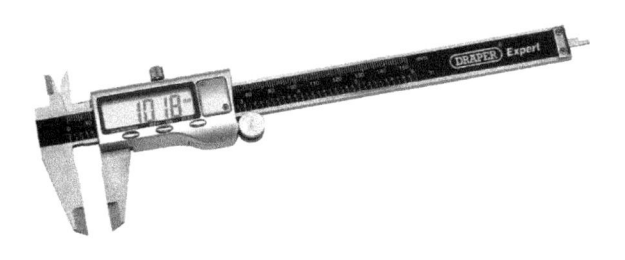

_____

_____

_____

_____

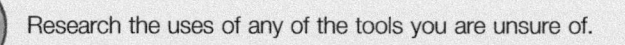

Research the uses of any of the tools you are unsure of.

Fabrication exercise

You may be studying for a qualification and the awarding body may stipulate that you have to provide evidence for fabricating an item and include measuring, drilling, cutting, filing and cutting threads.

Below is an example of a tool that you can fabricate, it is a dice screwdriver based on a design from a stubby screwdriver.

Dice screwdriver

On the following pages there are some engineer's drawings to make the tool; you would do well to draw the tool first using pencil and paper to practise the measuring techniques.

The engineer drawing is not to scale, what does this mean?

_____

_____

Here is a list of tools and materials needed to make the tool.

Materials

Low carbon steel block – to be cut 30 × 30 × 30mm and M6 steel bar.

Hand tools

A range of hand tools will be required including: rule, scribe, engineer's square, vice, hacksaw, flat file, half round file, drill, centre punch, 5mm drill bit, 6mm tap and die, thread lock and oxy acetylene bottles.

 Never use oxy acetylene unless fully trained to do so, if you are in doubt at any time during the process then ask your supervisor.

Fabricating the dice screwdriver

Use the following steps as a guide to fabricate the tool. Use the engineer's diagrams on page 54 and 55 to help you.

- **Obtain the materials.**
- **Cut the block to the correct dimensions.**
- **Dress the edges and smooth the corners.**
- **Cut the rod to the correct dimensions.**
- **Measure and mark the position of the dimples on the dice and centre punch.**
- **Drill the holes in the block using a 5mm drill bit. The dimples can be as deep as you wish but ensure they are an even depth for cosmetic value, around 1mm will be adequate. Number 1 on the dice should be drilled to a depth of 10mm to accept the screwdriver blade.**
- **Tap the hole in dice number 1 in the block to 6mm to create a thread. Use a correct lubricant when cutting the thread.**
- **Using a half round file, shape the screwdriver end to 5mm long and to a thickness of 1mm then dress sharp edges.**
- **Cut the thread on the bar to 6mm using a die to match the thread in the block, use the correct lubricant when cutting threads.**

- The screwdriver blade needs to be case hardened to make it stronger using oxy acetylene, you can do this with a variety of techniques depending on the material you are using. See the notes on the following page.

- Check all parts are made correctly and assemble by fastening the blade into the dice block using a thread lock material to prevent it coming loose.

 Research the process of case hardening and tempering and why it is used.

TIP The tool being made is for educational purposes. The quality of the finished product will depend on the materials used.

The screwdriver handle (dice block) can be made from low carbon steel (LCS) and the results will be good. The fact that it is low carbon steel will make the material relatively soft and easy to work with.

The screwdriver blade can be made from a variety of materials. Low carbon steel will be easy to work with and high carbon steel would make the tool longer lasting; however, it will be tougher to file and you will need a good quality die to cut the thread. High carbon steel is generally more expensive and more difficult to work with than low carbon steel.

Low carbon steel – case hardening

If you use LCS you will need to case harden the tip. What is the purpose of this?

_____

_____

How should LCS be case hardened?

_____

_____

_____

_____

High carbon steel (HCS) – case hardening

You will still need to case harden a screwdriver blade made from HCS but as it already has a high carbon content this will make the steel brittle, so it will need to be 'tempered'. How is tempering completed?

_____

_____

 Research the meaning of quenching, why it is necessary? Have a look at tempering colour charts.

Engineer's diagram of dice screwdriver

Material is LCS.

Dimensions are in mm.

Not to scale.

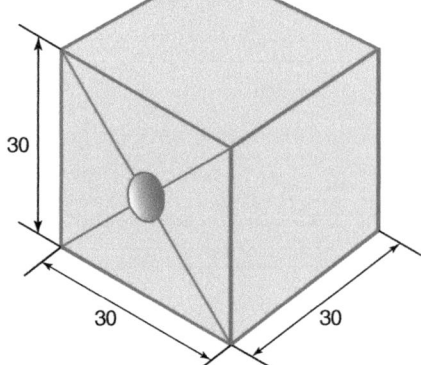

Dimples and number 1 hole to be drilled to 5mm diameter

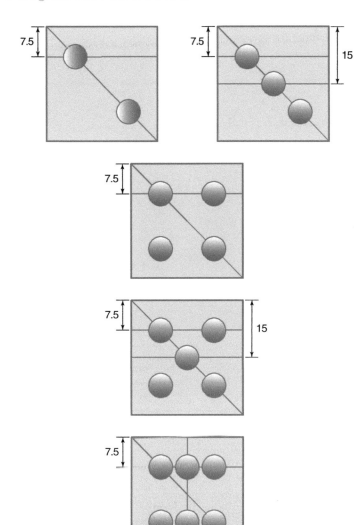

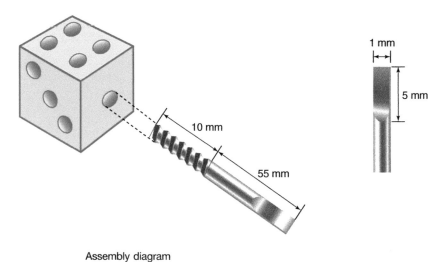

Assembly diagram

Screwdriver blade: M6 Bar.

Drill 5mm through number 1 hole 10mm deep.

Cut thread 6mm using a good quality tap and die and use the correct cutting fluid.

Screwdriver blade thread

 If in doubt at any time ask your supervisor.

Multiple choice questions

Choose the correct answer from a), b) or c) and place a tick [✓] after your answer.

1 Technician A says that a test lamp can be used to test a live airbag system, Technician B says that a power probe is the only tool that can be used, who is correct?

   a) Technician A [ ]

   b) Technician B [ ]

   c) Neither Technician A or B [ ]

2 During a cylinder leakage test, a blown head gasket is suspected; this is confirmed as air is located:

   a) In the coolant header tank [ ]

   b) In the air intake system [ ]

   c) In the exhaust system [ ]

3 The tool best used to test moisture content in brake fluid is:

   a) Hydrometer [ ]

   b) Hygrometer [ ]

   c) Thermometer [ ]

4 When case hardening you would add?

   a) Carbon powder [ ]

   b) Boron steel [ ]

   c) Adhesive [ ]

5 A multi-meter cannot be connected to a live circuit when measuring:

   a) Amps [ ]

   b) Volts [ ]

   c) Resistance [ ]

PART 3
ENGINES

USE THIS SPACE FOR LEARNER NOTES

_____

_____

_____

_____

_____

_____

_____

_____

SECTION 1
Diagnosis and rectification of engine mechanical related faults 59

1 Introduction to engine terms and conditions 60
2 Diagnosis of engine related faults 61
3 Lack of performance, misfiring and poor running 64
4 Multiple choice questions 73

SECTION 2
Diagnosis and rectification of engine management and emission related faults 74

1 Engine management systems 75
2 Air flow sensor (sometimes called a mass air flow meter) 75
3 Knock sensors 76
4 Coolant temperature sensor 77
5 MAP sensor 78
6 Camshaft and crankshaft sensors 79
7 Exhaust gas recirculation valves 80
8 Oxygen sensors (Lambda sensors, O_2 sensor) 81
9 Catalytic converters 83
10 NO_x sensor and catalyst 84
11 Diesel particulate filters (DPF) 85
12 Variable valve timing (VVT) 86
13 Turbo chargers 88
14 Stop start technology 90
15 Direct petrol fuel injection systems 90
16 Multiple choice questions 92

SECTION 3
Diagnosis and rectification of lubrication and cooling system related faults 93

1 Lubrication systems 94
2 Cooling systems 97
3 Multiple choice questions 99

SECTION 4
Overhauling engine units 100

1 Purpose and function requirements of engines 101
2 Timing belts and chains 102
3 Camshafts 103
4 Cylinder head removal and valve inspection 104
5 Valve stem oil seals 106
6 Replacing hydraulic lifters 107
7 Piston protrusion 107
8 Bent connecting rods 108
9 Piston rings 108
10 Cylinder bore inspection 109
11 Cylinder bore finish 109
12 Crankshaft end float 109
13 Crankshaft removal and inspection 110
14 Plastigauge 110
15 Multiple choice questions 112

SECTION 5
Hybrids and alternative fuels 113

1 Hybrid vehicles 114
2 Extended range/plug in hybrids 118
3 Electric vehicles 119
4 Hydrogen vehicles 121
5 Fuel cell vehicles 121
6 Multiple choice questions 122

SECTION 1

Diagnosis and rectification of engine mechanical related faults

USE THIS SPACE FOR LEARNER NOTES

_____

_____

Learning objectives

After studying this section you should be able to:

● Recognise engine terms and conditions.
● Understand methods of finding faults.

Key terms

Accumulator   A reservoir to store liquid.
Pinking/ping   A knocking noise caused by detonation (explosion) in the combustion chamber.
Misfiring   Failure of an explosion to occur in one or more cylinders while the engine is running; can be continuous or intermittent.

http://www.aa1car.com/library/engine_noise.htm

http://www.autotap.com/techlibrary/analyzing_ignition_misfires.asp

http://www.injectorcleaning.co.uk/index.htm

http://www.racq.com.au/motoring/cars/car_advice/car_fact_sheets/diesel_injector_cleaning

INTRODUCTION TO ENGINE TERMS AND CONDITIONS

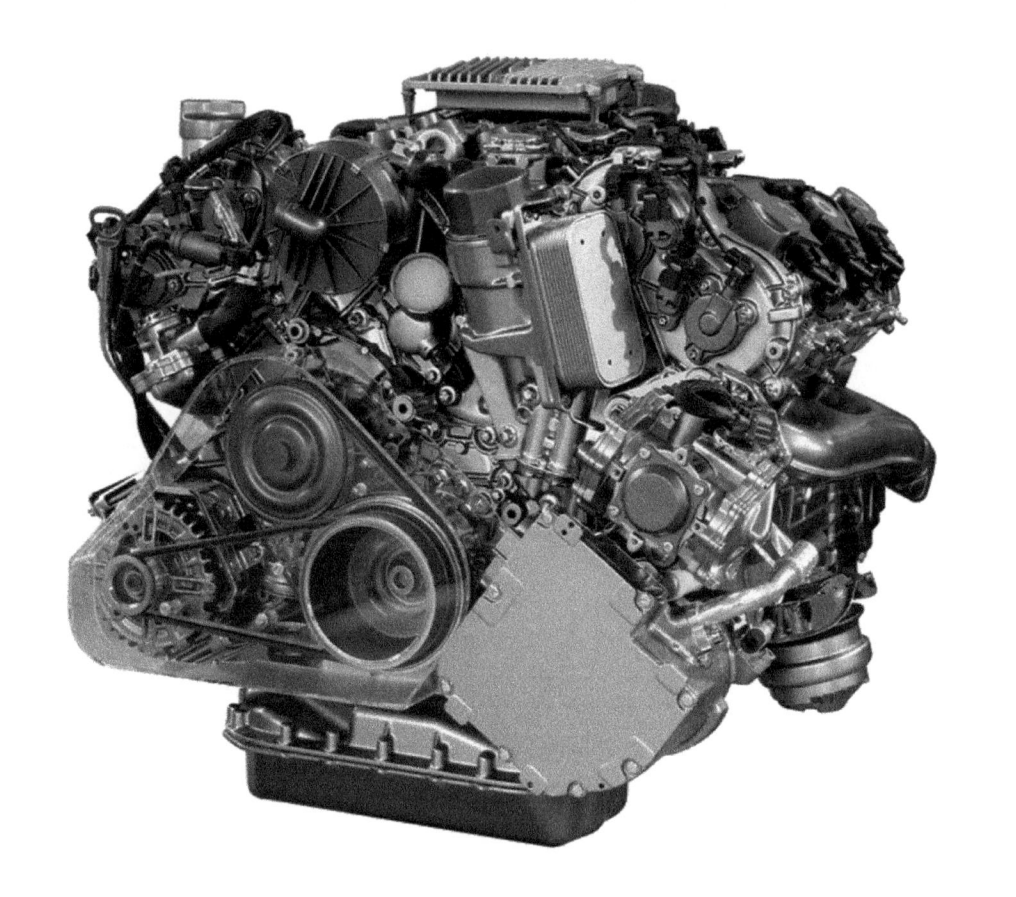

Whatever the type of engine or application, from the smallest around town car to the most powerful car, all engines work on the same principles.

For an engine to start and run, certain basic requirements must be present. List four such requirements:

1 _____

2 _____

3 _____

4 _____

If any one of the major factors in the list on the previous page are not present, the engine will stop.

With so few things needed to allow engines to run, why do they need so many technical sensors and parts?

_____

_____

_____

 If you are unfamiliar with major engine parts and operation then complete some research. Refer to *Vehicle Maintenance and Repair Level 1* and *Light Vehicle Maintenance and Repair Level 2* workbooks in the *Vehicle Maintenance* series.

Engine terms

Describe below the meanings of each of the terms given.

Torque – _____

Power – _____

Energy efficiency – _____

_____

Volumetric efficiency – _____

Thermal efficiency – _____

_____

Mechanical efficiency – _____

_____

Compression ratio – _____

_____

Cylinder swept volume – _____

Clearance volume – _____

_____

Honing – _____

Boring – _____

Reaming – _____

Valve lapping – _____

_____

Repairing engines can be physically demanding and often the tools used during repair can be dangerous if you are unfamiliar with them. Always be alert, concentrate carefully and ask your supervisor if you are unsure of what to do.

DIAGNOSIS OF ENGINE RELATED FAULTS

Noise

There are many jobs that need to be diagnosed due to noise, sometimes the noise can easily be diagnosed and may be caused by ancillaries attached to the engine such as water pumps and power-steering pumps, but often further investigations and equipment such as stethoscopes are needed.

Spark ignition engine

Pre-ignition, detonation, knock, ping or pinking, are all terms used when referring to spark ignition engines. All of the terms pretty much mean the same thing. Explain what the terms mean.

_____

_____

_____

_____

What could cause pre-ignition?

_____

_____

_____

When is pinking most likely to occur?

_____

_____

Are there any other components that can affect it?

_____

What can be done to stop pinking?

- _____

- _____

- _____

- _____

_____

Carbon on valves

 **TIP** Using the correct grade of unleaded fuel can prevent excess carbon.

Knock sensors

What is a knock sensor and what is its function in an engine?

_____

_____

_____

_____

A knock sensor close up

Label the diagram below using the following terms:

Valve **Fuel injector** **Cylinder wall**
knock sensor **Piston** **Coolant**

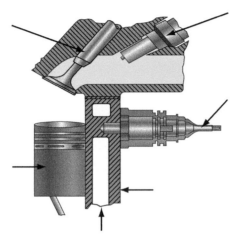

A knock sensor in situ

 CHECK If an engine is running low on oil and starting to rattle, this can confuse the knock sensor into thinking the engine is knocking, it may store a trouble code when really all that is needed is lubrication.

 Locate knock sensors on a variety of engines available in your place of study or workshop and make a note of their positions.

Engine noise may be produced by worn camshafts and sometimes hydraulic tappet faults.

Describe how to identify this type of noise.

_____

_____

_____

_____

CHECK If the engine has been standing for any length of time then the oil may have drained from hydraulic tappets. After running the engine for a few minutes they should quieten down, this is perfectly normal.

⚡ Take care near moving engine parts when diagnosing faults.

Hydraulic tappet

Worn crankshaft bearings may cause an engine to knock. In your workshop ask your supervisor to demonstrate this noise or search for videos of engine knock on the Internet. Describe the noise.

_____

_____

What is diesel knock?

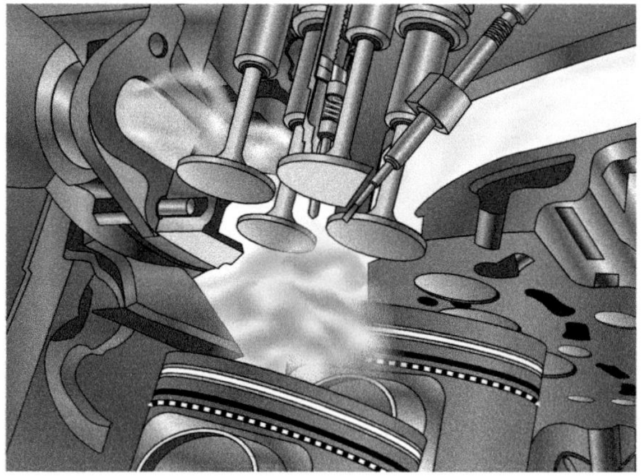

_____

_____

_____

An example of diesel knock

To summarise different engine noises, complete the table below to describe a possible fault for the symptom given.

Symptom	Possible cause
Rattling at the top end of the engine	
Pinking noise	

Symptom	Possible cause
Diesel knock	
Knocking from the lower engine	
Piston slap	

WWW **http://www.aa1car.com/library/engine_noise.htm**

LACK OF PERFORMANCE, MISFIRING AND POOR RUNNING

Compression test

An engine needs good compression to operate correctly; if the fuel and air is compressed and then 'leaked' out of the cylinder it can cause poor performance or a misfire.

List six examples that can cause low compression.

1 _____

2 _____

3 _____

4 _____

5 _____

6 _____

Compression test

Add the missing words to describe a compression test procedure:

throttle opening	**ignition**	**unrestricted**
spark plugs	**glow plug**	**diesel**

The engine fuel supply and _____ system should be disabled where possible. Remove

the _____ and insert the compression tester, ensure maximum _____ so there

is an _____ air flow into the engine. Crank the engine for a few seconds and record the

readings. On _____ engines follow manufacturer's recommendations. This may involve

fitting the compression tester into the injector hole or _____ hole.

A regular compression test may be considered a dry test, if compression on certain cylinders is found to be low, you may consider completing a wet test, what is meant be a wet test?

_____

_____

If the reading was to rise significantly during a wet test what may this indicate?

_____

_____

If two adjoining cylinders were low on compression what would you expect to find?

_____

_____

Cylinder leakage test

The compression test is ideal to diagnose engine faults before removing the cylinder head so you have an idea of what to look for in relation to the fault. Another good test is a cylinder leakage test.

Describe how to complete a cylinder leakage test and complete the diagram below to indicate how a cylinder leakage tester is connected to the engine.

_____

_____

_____

_____

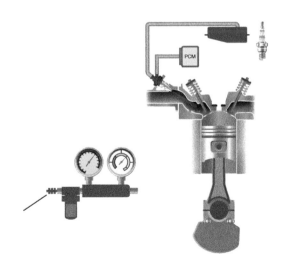

A cylinder leakage tester connected to the engine

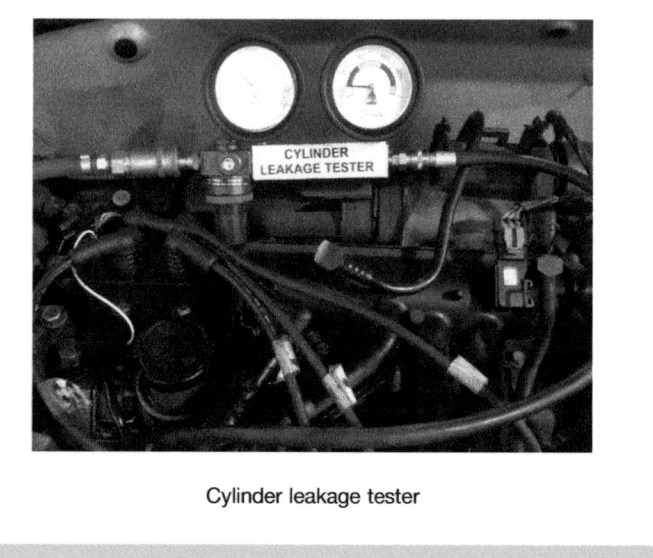

Cylinder leakage tester

 Take care when working with compressed air.

Complete the table below to describe the fault is relation to air loss.

Air leak	Most likely cause
Through the throttle body	
Through the dipstick tube	
Into the coolant header tank	

Air leak	Most likely cause
The air is heard escaping through the exhaust pipe	
The air is leaking from the spark plug hole of the cylinder next to the one being tested	

Cylinder power balance test

An engine needs all cylinders to contribute equally to its operation. A power balance test indicates cylinders that are not performing quite so well compared to other cylinders.

How is a power balance test completed?

_____

_____

_____

What temperature should the engine be when a power balance test is completed?

_____

Are there any risks to performing this particular test on a modern vehicle?

_____

_____

_____

Power balance test scenario

A six cylinder engine has a power balance test completed giving the following results.

Cylinder number	RPM drop
1	150RPM
2	150RPM
3	150RPM
4	150RPM
5	10RPM
6	20RPM

Using the table above, what is the most likely cause of the cylinders not contributing, and which other tests may be used in diagnosing the fault?

_____

_____

_____

There are other tests available to help diagnose engine faults including vacuum tests to check for faults such as blocked exhaust systems. Research engine vacuum tests to find out how they are completed. Make some revision notes below

Misfires

Misfires on older vehicles may have been checked in a similar way to a cylinder power balance test where the spark plugs would be disconnected in turn to see how it affects the cylinders. What are the dangers of performing a test in this way?

_____

_____

_____

With modern cars using coil packs and one coil per cylinder a misfire can best be detected using a code reader or what other instrument?

_____

If you are using a code reader you may find trouble codes stored in the ECU, you should check the basics such as wire connections and spark plugs. A code reader may also allow the use of an actuator test. What is an actuator test?

_____

_____

Spark plug faults

www http://www.autotap.com/techlibrary/analyzing_ignition_misfires.asp

Fuel problems – petrol

Low fuel pressures can cause a range of problems, for example poor starting or low power. One of the first tasks should be to check fuel pressure.

How may this test be completed?

_____

_____

_____

Complete some fuel pressure tests and compare readings with manufacturer's data.

Fuel pressure test	Reading 1	Reading 2	Reading 3

What may cause low fuel pressure?

_____

_____

Take care when disconnecting a fuel pressure tester as some fuel will probably leak out. If the engine is hot then this has obvious dangers.

The fuel rail will also have a fuel pressure regulator; basically this holds fuel in the fuel rail to ensure it is available for the next injector to operate. If the diaphragm was split in the fuel pressure regulator, which obvious fault would be present?

_____

_____

A fuel rail

Take care when testing fuel injectors and electrical connections, be aware of fire procedures.

Faulty petrol fuel injectors can cause misfires and poor running, they may need testing along with electrical connections.

Most petrol fuel injectors use a negative trigger – what does this mean?

_____

_____

_____

Injector connections can be tested with a multi-meter or voltmeter to determine correct operation. A 'Noid' lamp can also be used.

Research Noid lamps, find out what they do and how to use them.
What does a 'Noid' lamp check?

_____

_____

How are Noid lamps used?

_____

_____

_____

_____

Fuel injectors themselves can be tested using a code reader and often this will pinpoint individual injectors; however, following manufacturer's recommendations you could check the injector with an ohmmeter and compare the readings with other injectors on the same vehicle.

Injector testing with an ohmmeter

Fuel injectors can be tested and cleaned, often by specialist companies, and this can be less expensive than replacing injectors. The injectors are examined and flow rates are checked to examine their performance. They are cleaned using ultrasonic cleaners and then spray patterns are tested.

http://www.injectorcleaning.co.uk/index.htm

Never disconnect injector connections with the engine running, if the injector is open when disconnected there may be an uncontrollable rush of fuel which can destroy engines.

How do petrol injectors become contaminated and therefore faulty?

_____

_____

_____

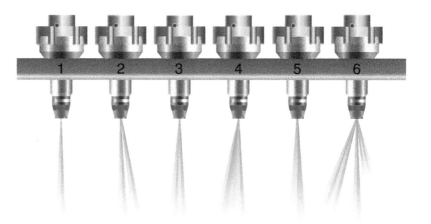

Injector spray patterns

Inspect the injector patterns on the previous page and describe each one.

1 _____

2 _____

3 _____

4 _____

5 _____

6 _____

Diesel injector testing

Diesel fuel injector faults

Diesel injectors can also fail–often the parts can wear.

A 'pop tester' is a popular method for specialists to check injectors. Name three tests that are completed during a pop test.

1 _____

2 _____

3 _____

Testing injectors can be extremely dangerous, especially modern common rail diesel injectors due to very high operating pressures. If in doubt see your supervisor and where possible use specialist testing companies.

Research injector testing and in particular 'pop testing'. Consider the dangers involved in testing injectors and discuss in small groups.

The image to the right shows a 'pop tester'.
Why might the fuel injector leak fuel under pressure and what affect would this have on the engine?

_____

_____

_____

_____

_____

Diesel injector testing

Diagnosing common rail diesel injection faults

Most modern diesel engines are 'common rail' injection which work under extreme pressures, 26,000 psi is not uncommon. Why have manufacturers started to use common rail injection systems?

_____

_____

Convert 26,000 psi into bar: _____

Consider a car tyre which on average is inflated to a pressure of 30psi which equals 2 bar.

What is the purpose of the common rail?

_____

_____

Fuel rail

Due to the high pressures of the fuel system list five major safety factors to consider:

1 _____

2 _____

3 _____

4 _____

5 _____

Check manufacturer's recommendations when removing parts from the system. Some pipes are considered single use only as you cannot guarantee a reliable seal once opened. Also individual injectors may be programmed to the ECU; don't mix them up if removed. If individual injectors are replaced, re-programming or calibration may be required.

 http://www.racq.com.au/motoring/cars/car_advice/car_fact_sheets/ diesel_injector_cleaning

Common rail injection systems use an ECU to operate. List nine parameters the ECU may be checking:

- _____
- _____
- _____
- _____
- _____
- _____
- _____
- _____
- _____

When diagnosing common rail faults, what would be a good diagnostic starting point?

_____

_____

_____

When working on common rail injection faults you should always consider checking low and high fuel pressure circuits. Which part of the system could be considered the low pressure fuel system?

_____

_____

What would you be looking for on inspection of the low pressure circuit?

_____

_____

Which components are considered to be part of the high pressure system?

_____

_____

CHECK A sensor is commonly incorporated into the system and high pressures can normally be read using a good diagnostic scan tool.

Some manufacturers may recommend the use of a pressure testing kit. This tool is particularly useful when checking fuel pressures on start up.

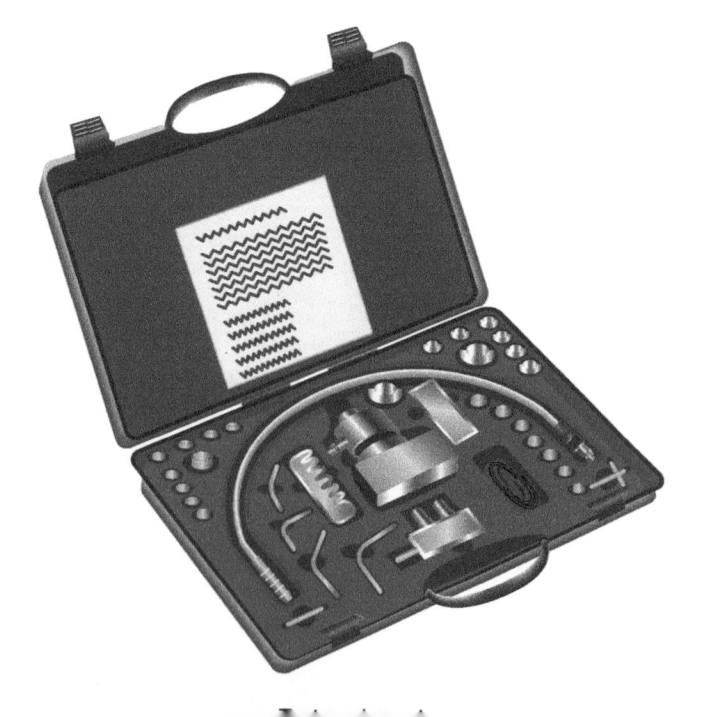

This test equipment should be fitted between which components?

_____

_____

_____

Another method of checking diesel injector performance is to complete a fuel leak off quantity test. Describe how this test is done and the equipment needed.

_____

_____

_____

_____

Injector leak back test

Multiple choice questions

Choose the correct answer from a), b) or c) and place a tick [✓] after your answer.

1 **A vehicle lacks power, Technician A says the best diagnostic tool will be a cylinder leakage tester, Technician B says the car should be checked with a diagnostic tester, who is correct?**

   a) Technician A [ ]

   b) Technician B [ ]

   c) Both Technician A and Technician B [ ]

2 **The test procedure to check the operating pressure and spray pattern of a diesel injector is known as?**

   a) A 'pop' test [ ]

   b) A 'burst' test [ ]

   c) A 'compression' test [ ]

3 **When completing a compression test it may be advised that you then complete a 'wet test'. To do this what should be added to the cylinder?**

   a) A small quantity of petrol [ ]

   b) A small quantity of oil [ ]

   c) A small quantity of water [ ]

4 **The meaning of torque is –**

   a) The twisting or turning force [ ]

   b) The number of turns [ ]

   c) How loose the component is [ ]

5 **On a common rail diesel injection system, which component would be considered a low pressure component?**

   a) The common rail [ ]

   b) The pipe from the fuel tank to the pump [ ]

   c) The pipe from the pump to the fuel rail [ ]

SECTION 2

Diagnosis and rectification of engine management and emission related faults

USE THIS SPACE FOR LEARNER NOTES

_____

_____

Learning objectives

After studying this section you should be able to:

- **Recognise engine management systems and related faults.**
- **Understand emission related faults.**
- **Understand methods of finding faults.**

Key terms

Electronic control unit (ECU) An on-board vehicle system computer.
Injector pulse width The time that a fuel injector stays open.
Rectification To repair something.
Lean An engine that is running on a 'more air/less fuel' content.
Rich An engine that is running on a 'more fuel/less air' content.

http://www.picoauto.com

http://easyautodiagnostics.com

http://www.ngk-elearning.com

http://www.carbibles.com/fuel_engine_bible_vvt.html

ENGINE MANAGEMENT SYSTEMS

CHECK Engine management systems are obviously monitored by an electronic control unit (ECU). The ECU has maps programmed into it and sensors are continually monitored, the ECU has the ability to change specifications with reference to given inputs. Should a sensor start to work outside of its correct limits then the engine management light or check light will be illuminated warning the driver of a fault, The system may activate 'limp home' mode where the ECU can estimate the correct values.

AIR FLOW SENSOR (SOMETIMES CALLED A MASS AIR FLOW METER)

Air flow meter

What is the purpose of an air flow meter?

_____

_____

_____

Which is the most common type of air flow meter and what does it measure?

_____

_____

How does this meter work?

_____

_____

_____

_____

_____

List four common air flow meter related faults:

- _____

_____

_____

- _____

_____

- _____

_____

- _____

_____

 CHECK Air flow meters are best diagnosed using code readers and oscilloscope readings. There are cleaning agents available to clean the hot wire but always follow the manufacturer's instructions to avoid damage to the sensor.

Blocked throttle body

It is important that an air flow meter is connected in the correct manner. Using the picture below explain how to identify how the meter should be fitted.

Air flow meter

KNOCK SENSORS

Where is a knock sensor most likely to be located on the engine?

The sensor detects pre-ignition or engine knock; it sends a signal to the ECU to adjust ignition timing until knock is controlled. If uncontrolled then major engine damage can occur.

Melted piston and knock sensor

A knock sensor is a 'piezoelectric' device, what does this term mean?

Oscilloscope reading

Reproducing a knock sensor fault may be difficult as at idle in the workshop the engine may be running fine. How can you quickly check a knock sensor?

COOLANT TEMPERATURE SENSOR

A coolant temperature sensor is often referred to as a thermistor; this is basically a resistor which varies the value of its output in accordance to temperature changes. Commonly a NTC (negative thermal co-efficient) sensor is used.

When working with NTC sensors what happens to the resistance value as the temperature increases?

The sensor can be tested using a scan tool and if available read live data from the sensor. Alternatively you could disconnect the sensor and check the resistance across the two terminals as the engine warms up.

Testing a coolant temperature sensor

When measuring ohms resistance as the engine warms up, be aware and keep test equipment and yourself clear of moving engine parts.

Enter the workshop in your place of training or study. Using an ohmmeter and an NTC coolant thermometer plot a graph below to show the decrease in resistance as temperature rises.

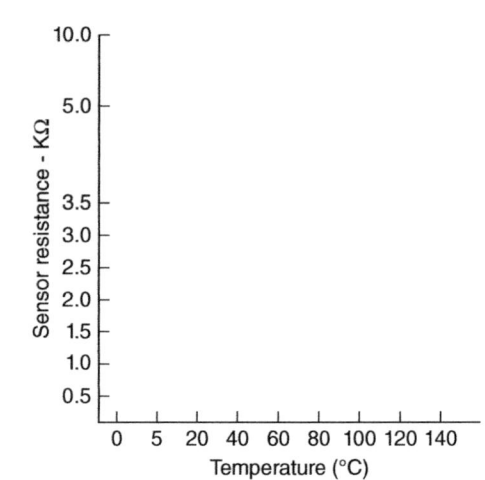

Coolant temperature sensor

MAP SENSOR

What does MAP sensor stand for?

What does a MAP sensor do?

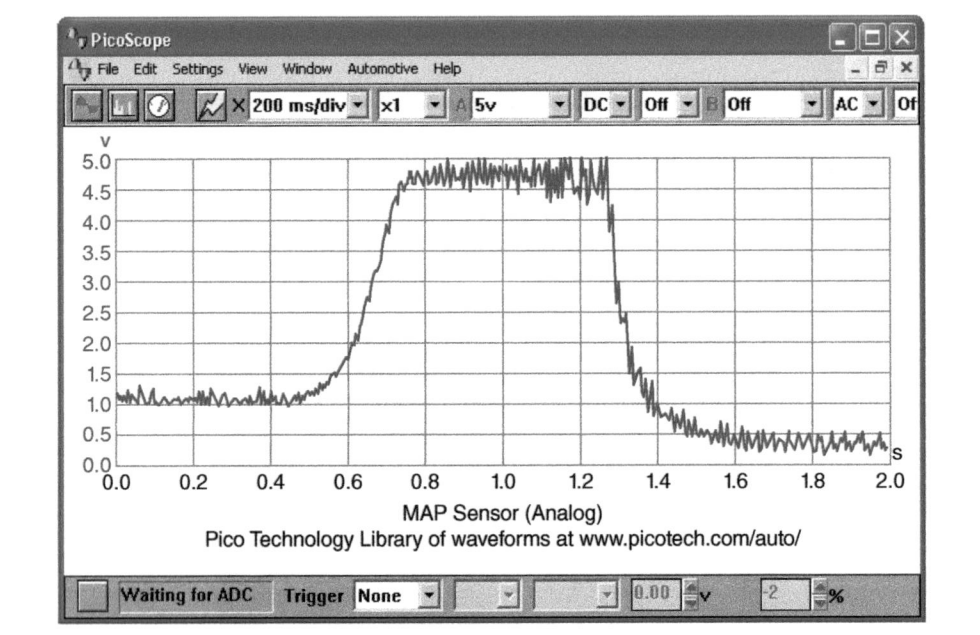

Oscilloscope pattern

The image above shows an oscilloscope waveform of a typical MAP sensor. Explain what is happening using the waveform.

How may incorrect voltages affect the running of the vehicle?

Following manufacturer's recommendations research how to diagnose a MAP sensor using a voltmeter and a vacuum gauge.

As well as checking the MAP sensor, check for split vacuum pipes and corroded wiring connections.

CAMSHAFT AND CRANKSHAFT SENSORS

The camshaft and crankshaft sensors have two common purposes, why are these sensors fitted to modern engines?

1 _____

2 _____

The images below show a crankshaft and camshaft sensor. Which one is which?

Engine sensors

Engine sensors

Camshaft and crankshaft sensors work in a similar way to each other, there are two types of sensor that manufacturers use – analogue and digital; how can you recognise the different types?

Analogue (Inductive type)

_____

_____

Digital (Hall sensor)

_____

_____

_____

What advantages do digital sensors have over analogue sensors?

_____

_____

_____

List four ways to test an analogue (Inductive) crankshaft sensor.

1 _____

2 _____

3 _____

_____

4 _____

_____

_____

 When carrying out tests keep equipment and yourself away from moving engine parts, if in doubt ask your supervisor. Ensure no one can start the engine until you are ready to start the test.

www http://easyautodiagnostics.com

Describe how to test a digital (Hall) crankshaft sensor

_____

_____

_____

List four symptoms of faulty crankshaft or camshaft sensors:

1 _____

2 _____

3 _____

4 _____

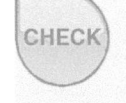 CHECK It is important that these sensors are correctly fitted; you may be able to adjust the air gap on a camshaft sensor but this will affect its performance if incorrect. Loose timing belts and stretched timing chains can also affect the performance of these sensors, check these out during diagnosis.

EXHAUST GAS RECIRCULATION VALVES

An exhaust gas recirculation valve forms part of the emission control system on a vehicle and as the name suggests recirculates some of the exhaust gases. Since the gases have already been burnt, how can recirculating the gases help with emission control?

_____

_____

_____

Smog

_____

_____

Why is it common for exhaust gas recirculation valves to fail?

_____

_____

What may a customer complain of which may lead you to diagnosing the exhaust gas recirculation valve?

_____

How can an exhaust gas recirculation valve be checked?

_____

_____

_____

 Research different types of exhaust gas recirculation valves. Find out the prices, makes and features of different exhaust gas recirculation valves.

OXYGEN SENSORS (LAMBDA SENSORS, O₂ SENSOR)

Technicians refer to these sensors using any of the above terms.

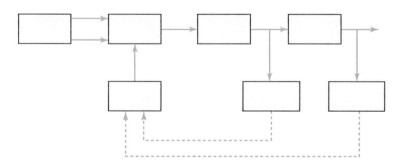

Add the following terms to the block diagram above to demonstrate a closed loop control system using Lambda sensors:

engine	**fuel/air delivery**	**fuel/air**
upstream O₂ sensor	**ECU**	
downstream O₂ sensor	**catalytic converter**	

 Research the different terms used to identify lambda sensor positions in the exhaust system. Write them down as revision notes.

 Catalytic converters can become very hot in operation.

Catalytic converters are used on modern vehicles to control emissions, an oxygen sensor is fitted upstream of the catalytic converter to relay information to the ECU to ensure excess fuel is not burnt in the engine. Why is an oxygen sensor also fitted downstream of the catalytic converter?

_____

What is meant by the stoichiometric air/fuel ratio, and what is the connection to a lambda sensor?

_____

_____

_____

There are currently three popular types of oxygen sensor:

1 **Zirconia type**

2 **Titania type**

3 **UEGO (Universal Exhaust Gas Oxygen) type**

Zirconia type lambda sensor

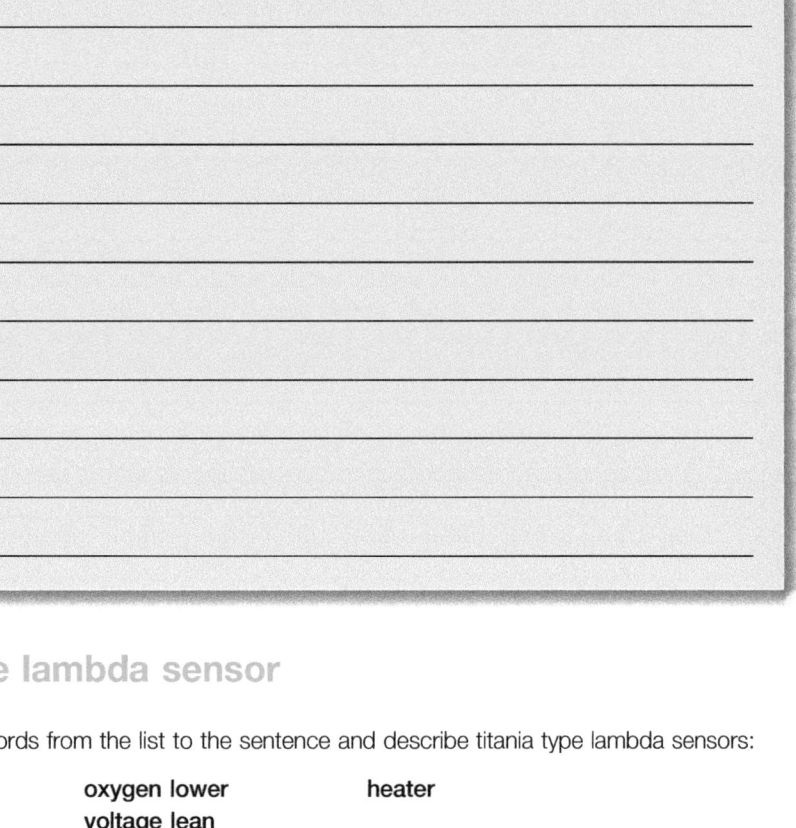

Visit this website **http://www.ngk-elearning.com**. Research Zirconia type sensors and write six facts about them below.

1 _____

2 _____

_____

3 _____

_____

4 _____

_____

5 _____

_____

6 _____

_____

Titania type lambda sensor

Add the missing words from the list to the sentence and describe titania type lambda sensors:

resistance   **oxygen lower**   **heater**
rich   **voltage lean**

The titania sensor looks similar in design to the zirconia sensor; these sensors do not generate

_____ as the zirconia type but electrical _____ which changes in relation to the _____

content in the exhaust gases. Excess oxygen (_____mixture) causes the resistance to rise; low

oxygen (_____ mixture) causes the resistance to _____. There is no need for a pocket of

ambient air as a reference so the sensor is generally smaller. These sensors generally have a

_____ circuit.

UEGO (Universal Exhaust Gas Oxygen) type lambda sensors

These sensors are generally identified by the fact they have five wires and they always contain a heater circuit. This sensor does not need a reference of ambient air. The sensor has a small tube, or detection chamber which is exposed to exhaust gas. The sensor uses a zirconium element and tries to keep the exhaust gas as close to stoichiometric air fuel ratio as possible by pumping oxygen in or out of the chamber. The value of the pumping current corresponds to the air fuel ratio of the exhaust gas.

Oxygen sensor

Why are these sensors used in Formula 1 vehicles now appearing on road cars?

_____

Why do lambda sensors use a heater circuit?

_____

Oxygen sensor socket

An oxygen sensor socket should be used for correct replacement and to take care of hot parts.

Diagnosis of lambda sensors

The heater element could easily be checked using which tool?

_____

_____

_____

Diagnostics can also be completed using a diagnostic tester and oscilloscope.

CATALYTIC CONVERTERS

Catalytic converters clean dangerous exhaust gases into less harmful gases. This exhaust box contains a ceramic element coated in a catalyst.

Give a definition of a catalyst.

_____

_____

Add to the diagram below the gases that enter and leave the catalytic converter on a petrol engine.

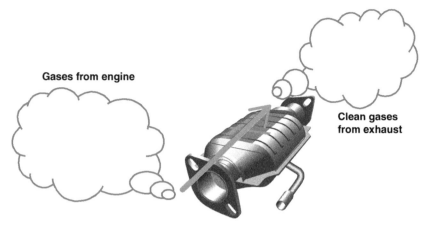

Gases from engine

Clean gases from exhaust

As the gases pass through the catalytic converter catalyst a chemical reaction takes place and harmful gases are changed to less harmful gases.

Catalyst

Label parts of the catalytic converter on the diagram below using the terms from this list:

stainless steel housing	**carbon dioxide**	**nitrogen oxide**
ceramic monolith	**nitrogen**	**hydrocarbons**
expanding mat	**water**	
lambda probe	**carbon monoxide**	

Explain the purpose of the items numbered on the diagram:

1 _____

2 _____

1.

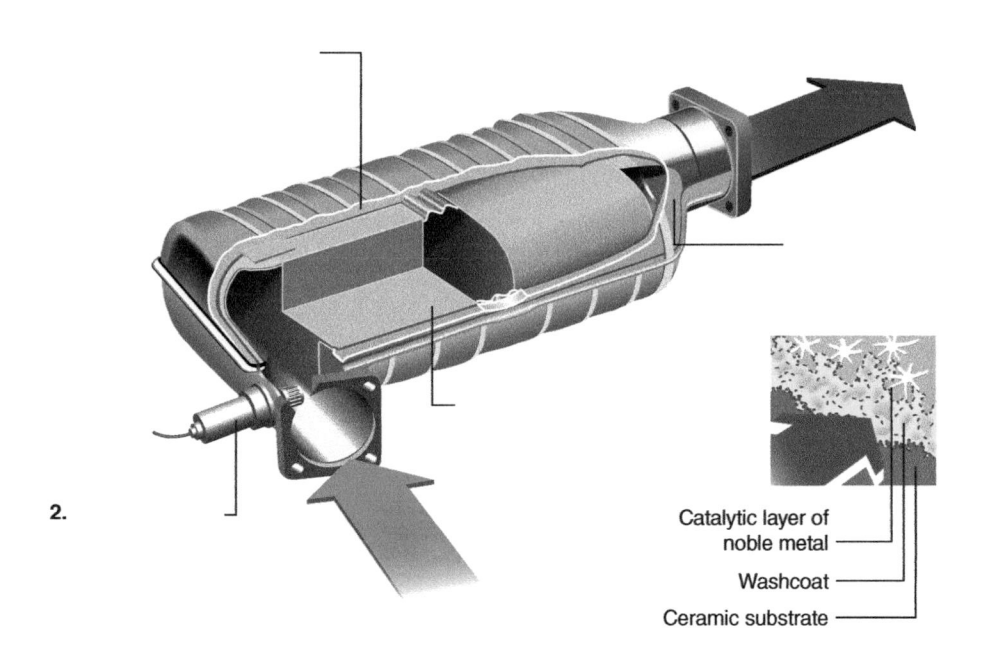

Catalytic layer of
noble metal _____

Washcoat _____

Ceramic substrate _____

2.

The catalytic converter only works correctly when it is hot. Describe what could happen if the vehicle had a misfire.

_____

_____

_____

Research diesel catalytic converters and the emissions converted whilst in use.

NOₓ SENSOR AND CATALYST

Manufacturers are now developing **lean** burn engines to meet strict emission standards. These may include direct injection petrol engines where the fuel injector is inside the cylinder, similar in operation to a diesel, allowing more than one delivery of fuel per cycle which is very effective in emission control.

These engines run very lean, what affect does this have on emissions?

_____

_____

How can manufacturers overcome this situation?

_____

_____

CHECK Note lean burn engines still use a traditional three-way catalytic converter to reduce carbon monoxide and hydrocarbons.

Briefly describe how a NOₓ catalytic converter and sensor work.

_____

_____

_____

Why are catalytic converters fitted as close as possible to the exhaust manifold?

_____

_____

_____

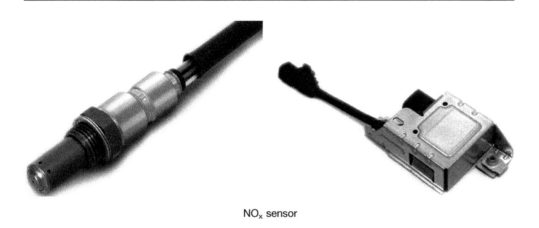

NO$_x$ sensor

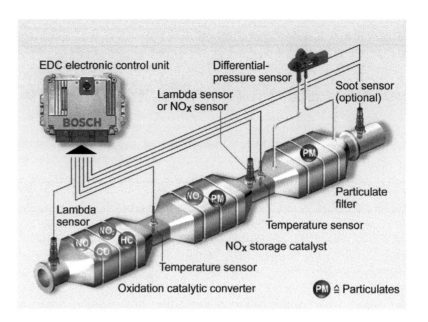

Diesel particulate filter operation

DIESEL PARTICULATE FILTERS (DPF)

Diesel engines have been subject to emission control in a similar way to petrol engines over the last few years. Manufacturers have developed common rail injection systems to better control fuel, exhaust gas recirculation has been introduced to reduce NO$_x$ but diesel fuel still produces soot (carbon deposits) when burnt, this has brought about the introduction of diesel particulate filters (DPF).

Diesel particulate filter operation

Two types of DPF are commonly fitted, one that uses an additive and one that doesn't. Explain the operation of each below and using the space provided on page 86.

Without additive

_____

_____

_____

With additive

_____

_____

_____

_____

> **CHECK** Diesel cars may be unsuitable for drivers that only complete short journeys, as the DPF needs certain conditions to operate such as long motorway journeys.

Diagnosis of catalytic converters and diesel particulate filters

List some ways to check for faults:

- _____
- _____
  _____
- _____
  _____
- _____
  _____
- _____
  _____
- _____
  _____

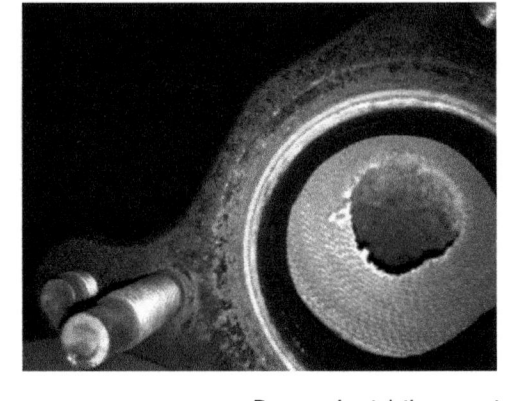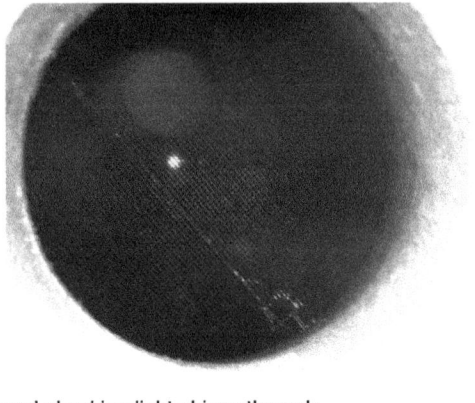

Damaged catalytic converter and checking light shines through

VARIABLE VALVE TIMING (VVT)

In order to improve engine performance and increase fuel efficiency, many manufacturers are now using variable valve timing. There are many different systems and operating mechanisms; however, the principle is the same. What is the main goal of variable valve timing?

_____

_____

_____

The ECU will need data relating to crankshaft and camshaft positions and speed, along with the throttle position sensor in order to operate the variable valve timing mechanism.

> **TIP** Take care when removing camshafts and VVT systems and ensure correct installation.

There are many different arrangements for operating the valves, some systems only operate inlet valves but many operate the inlet and exhaust valves.

More information on operating valves and some animations can be found here:
http://www.carbibles.com/fuel_engine_bible_vvt.html

The system below clearly demonstrates how one manufacturer alters the valve opening period.

Variable valve timing

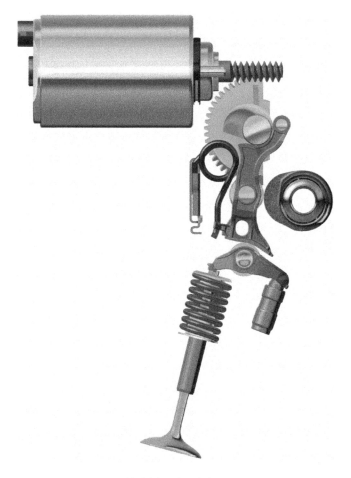

Variable valve timing

TURBO CHARGERS

It is becoming common for manufacturers to reduce engine sizes to save weight, space and cost.

Smaller engines fitted with turbo chargers can offer almost the same output as some larger engines but with greater fuel efficiency

Briefly explain how a turbo charger works.

_____

_____

_____

_____

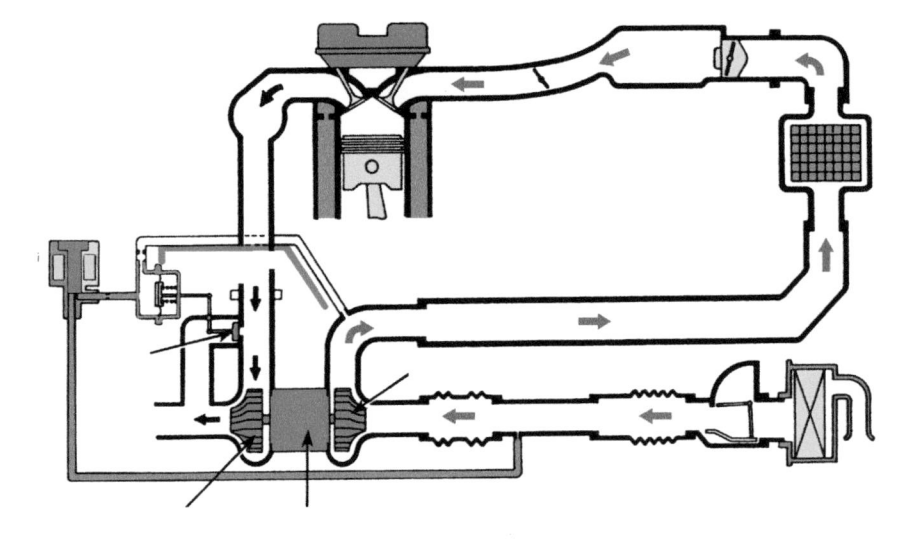

Turbo charger

Add the following labels to the above diagram:

air cleaner	turbo charger	turbo charger vacuum
air flow meter	intercooler	switching valve
compressor wheel	waste-gate valve	
turbine wheel	waste gate actuator	

What is the purpose of the waste-gate valve?

_____

_____

_____

_____

_____

Explain the purpose of a dump valve.

_____

_____

_____

_____

_____

What is another name for a dump valve?

_____

Are there any disadvantages to venting the air into the atmosphere?

_____

_____

_____

What alternative sensor can be used instead of an airflow meter and why?

_____

_____

CHECK   Without a dump valve, when the throttle is closed the pressurised air would try to flow back the way it came, through the compressor. This could place unnecessary load onto the compressor.

Why is an intercooler fitted?

_____

_____

_____

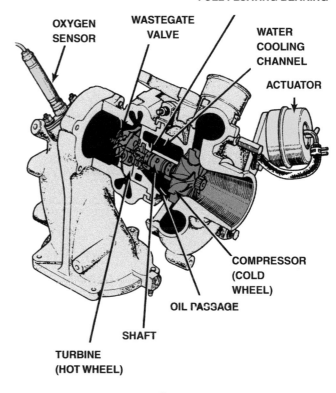

Turbo

Turbo charger inspection

List six simple ways to determine turbo charger related faults using the space below and on opposite column:

- _____
- _____
- _____
- _____
- _____
- _____

STOP START TECHNOLOGY

A stop start system allows for a reduction in fuel emissions and improves fuel economy.

What is the basic operating principle of stop start technology?

_____

_____

List eight situations where stop start technology will NOT work:

1 _____

2 _____

3 _____

4 _____

5 _____

6 _____

7 _____

8 _____

TIP Before diagnosing faults ensure the customer knows how to operate the system— there may not actually be a fault.

A vehicle that uses stop start technology needs to have certain components modified or upgraded over those in a conventional vehicle.

Add three more to the points mentioned below:

1 If the vehicle is constantly stop/starting in slow traffic then a large capacity battery will be needed.

2 _____

_____

3 _____

4 _____

_____

_____

DIRECT PETROL FUEL INJECTION SYSTEMS

Direct petrol injection relates to how technology has been developed to try and further improve economy and emissions.

On a conventional multi-point fuel injection system where would the injectors have been situated?

_____

The direct injection system uses a high pressure pump and fuel rail, the injectors are electronically controlled and inject fuel directly into the cylinders; this system is very similar in operation to which other vehicle operating system?

_____

_____

A direct petrol injection system can be designed to run without a throttle plate in the throttle body. How can this be an advantage?

_____

_____

_____

_____

The amount of fuel delivered can be very accurate and will be affected by the amount of time that the injector is opened, what is this term known as?

_____

_____

It is common to find that the piston crown is shaped to encourage the fuel and air to swirl; this helps to mix the fuel and air together. The engine management system can accurately inject the fuel as required. Depending on the operating conditions the fuel delivery can be altered to before or after the end of the compression stroke.

There are generally two injection modes. Explain each mode.

Stratified

_____

_____

_____

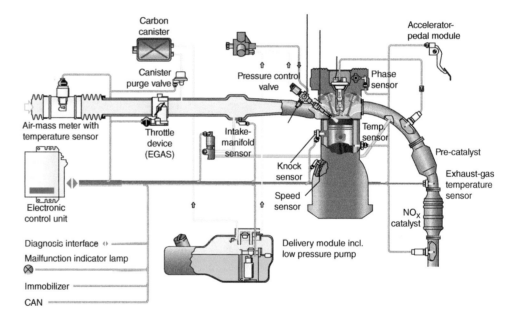

Direct injection parts

Homogenous

_____

_____

Add the following labels to the diagram of the 'Direct petrol injection system' below.

EGR valve   Injection coil   High pressure pump

Fuel rail   Fuel pressure sensor

Injector   Oxygen sensor (x2)

Direct petrol injection system

Take care when working on high pressure fuel systems.

Why does the system need to work under such high pressures?

_____

_____

_____

Due to the lean operating conditions the engine uses NO_x catalysts and EGR valves.

The diagrams below show stratified and homogenous operations. Identify which diagram is illustrating stratified and homogenous mode?

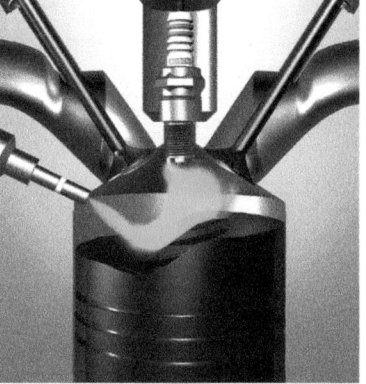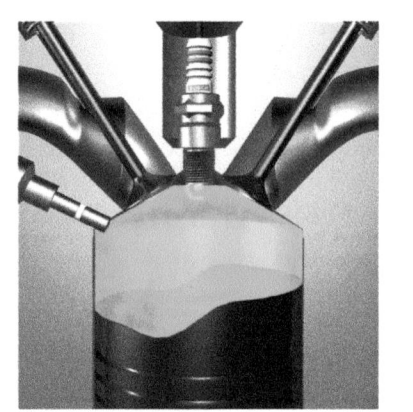

Stratified and homogenous operations

Multiple choice questions

Choose the correct answer from a), b) or c) and place a tick [✓] after your answer.

1 **What is a MAP sensor?**

a) Manifold Absolute Pressure sensor [ ]

b) Manifold Average Pressure sensor [ ]

c) Manifold Altitude Pressure sensor [ ]

2 **Where is a knock sensor most likely to be located on the engine?**

a) The fly wheel [ ]

b) Between the camshaft pulleys [ ]

c) On the engine block [ ]

3 **When testing a coolant temperature sensor what happens to the resistance as the temperature rises?**

a) it remains the same [ ]

b) It increases [ ]

c) It decreases [ ]

4 **Direct petrol injection systems are similar in operation to which other system?**

a) Common rail diesel engines [ ]

b) Single point fuel injection systems [ ]

c) Naturally aspirated engines [ ]

5 **An EGR valve may be incorporated to reduce?**

a) Nitrogen oxides [ ]

b) Nitrogen carbides [ ]

c) Nitrogen dioxide [ ]

SECTION 3

Diagnosis and rectification of lubrication and cooling system related faults

Learning objectives

After studying this section you should be able to:

- Recognise engine lubrication system faults.
- Understand cooling related faults.
- Understand methods of finding faults.

Key terms

Oil consumption   The use of oil by an engine.
Oil gallery   A passageway in the engine which allows oil to flow to other areas of the engine.

 WWW   http://www.ringautomotive.co.uk

http://www.carbibles.com/engineoil_bible.html

http://www.commaoil.com

USE THIS SPACE FOR LEARNER NOTES

_____

_____

LUBRICATION SYSTEMS

When working on lubrication systems ensure you wear correct PPE to prevent unnecessary contact with the oil.

The Society of Automotive Engineers (SAE) has established an oil viscosity system that is accepted throughout the motor industry. What is meant by the term viscosity?

_____

Research oil viscosities and classifications, find out about different oil for different engine types.

WWW **http://www.commaoil.com**

Oil consumption

All engines use some oil by design, and some engines will develop oil leaks, so it is important that oil consumption is not excessive or engine damage can occur. Oil leaks can occur through damaged or split gaskets and seals but can be common if the crankcase pressure becomes too high; how can this happen?

_____

_____

If a customer complains that the engine is using oil it may be necessary to do an oil consumption test, how can this be completed?

_____

_____

_____

A customer has arrived at the garage and has complained of an oil leak.

Discuss in small groups how you would locate the part that was leaking oil, record your answer below.

● _____

● _____

List all of the parts on the engine that could cause an oil leak.

- _____
- _____
- _____
- _____
- _____
- _____
- _____
- _____
- _____
- _____
- _____
- _____
- _____
- _____
- _____
- _____

Oil pumps

Oil pumps are commonly rotary vane type pumps; however, there are other designs available such as a crescent type pump or a simple gear type pump. The oil pump generally has a relief valve which prevents excess pressure building up at high engine speeds.

Label the diagram below which shows an oil pump.

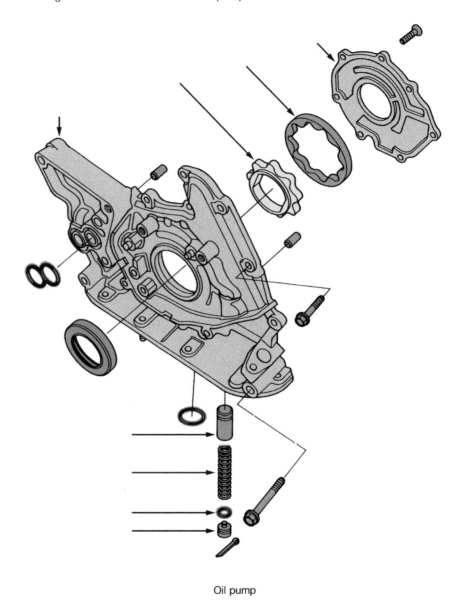

Oil pump

When an engine is rebuilt, the oil pump should be inspected for wear and damage–although the oil pump is probably the best lubricated part in the engine, the oil passing through the pump is unfiltered.

Explain the diagrams below. Describe what is taking place in each diagram.

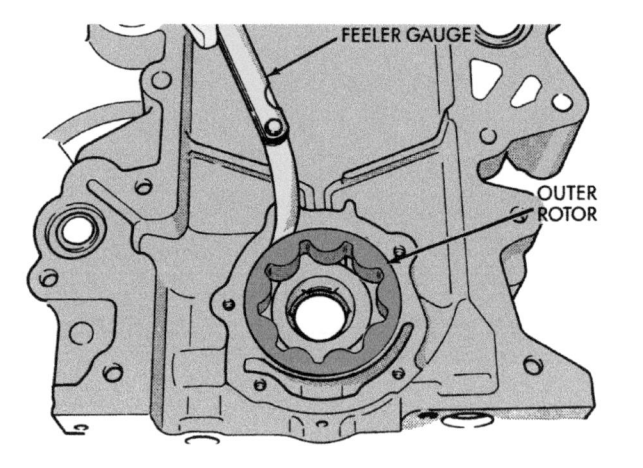

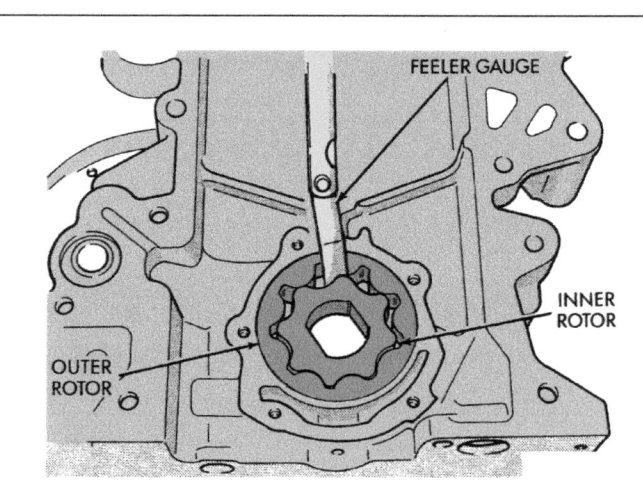

The performance of the oil pump and the clearances through which the oil flows can be determined by completing an oil pressure test.

Using the diagram below, explain how to complete an oil pressure test.

_____

_____

_____

Oil pressure test

TIP Oil pressure readings can be too high as well as too low.

What might a customer complain of that would make you decide to complete an oil pressure test?

- _____
- _____
- _____
- _____

CHECK A rattling camshaft is a classic sign that oil pressure is low and not enough oil is reaching the top of the engine to lubricate parts. Do not get confused with a possible sticking hydraulic lifter which will make a similar noise.

 When the engine is at operating temperature the oil will be hot and could cause severe burns, be careful when installing and removing testing equipment.

COOLING SYSTEMS

The internal combustion engine generates excessive amounts of heat and needs liquid cooling systems to keep it at an operating temperature. The temperature needs to be controlled as engines are more efficient at operating temperature but must not be allowed to overheat.

Thermostat

The thermostat controls the temperature and amount of coolant passing to the radiator. Commonly they have a pellet of wax in them, how do they work?

_____
_____
_____

 Always dispose of waste coolant in a responsible manner.

Testing thermostats

What four effects may a faulty thermostat have on an engine?

1 _____
2 _____
3 _____
4 _____

Explain a simple method to test thermostat operation without removal.

_____
_____
_____
_____

Explain a method to test a thermostat that has been removed.

_____
_____
_____
_____

Water pump

A water pump circulates coolant around the engine and to the radiator to cool the engine.

Name three ways that a water pump can fail:

1 _____

2 _____

3 _____

If the vehicle has a coolant leak an experienced technician may firstly recognise the smell of the coolant, upon checking a visual inspection may be enough to locate a leak.

Name two other recommended ways to diagnose coolant leaks.

1 _____

2 _____

Describe how fluorescent dye can be used to diagnose a coolant leak.

_____

_____

_____

_____

_____

_____

Coolant leak detector dye

WWW http://www.ringautomotive.co.uk

The tool shown to the right is a coolant pressure tester, how is it used to diagnose cooling system faults?

_____

_____

_____

_____

_____

_____

_____

Coolant pressure tester

Hot coolant can cause severe scalding and is normally under pressure when hot – take care!

TIP Should a coolant leak not be visible check the interior of the vehicle to see if the heater matrix is leaking, this may also be identified by a coolant smell in the car or a damp carpet.

In your place of training or study locate a coolant pressure tester and using the manufacturer's instructions pressure test a cooling system and pressure cap.

Scenario

A customer arrives at your garage and says that their temperature gauge is reading high but they have not checked for any signs of leaks.

In small groups decide how to set about diagnosing the fault and record your answers below.

_____

_____

_____

_____

_____

_____

_____

_____

_____

_____

_____

_____

_____

Multiple choice questions

Choose the correct answer from a), b) or c) and place a tick [✓] after your answer.

1 **The job of the cooling system is to?**

   a) Keep the engine as cool as possible [ ]

   b) Allow the engine to get to operating temperature and keep it at a temperature to ensure engine efficiency [ ]

   c) Allow the engine to get as hot as possible to ensure good heating for the passengers [ ]

2 **If an engine has a weak anti-freeze mixture and a customer complains that the car heater is poor in operation a suspected fault could be?**

   a) The water pump impeller has broken from the shaft [ ]

   b) The heater motor has frozen up [ ]

   c) Ice is preventing the thermostat from closing [ ]

3 **Why might an oil pump include a pressure relief valve?**

   a) To prevent the pump running too quickly [ ]

   b) To ensure maximum pressure is always supplied [ ]

   c) To return oil to the sump if pressure becomes too high [ ]

4 **A customer complains of a rattling engine, the first action to be taken by the technician should be?**

   a) To carry out an oil pressure test [ ]

   b) To check the oil pump for operation [ ]

   c) To check and if necessary correct the oil level [ ]

5 **Should an engine leak either oil or coolant a dye can be added, what is the purpose of the dye?**

   a) To locate the leak using an ultraviolet light [ ]

   b) To seal the leak [ ]

   c) To locate the leak using a leak detecting paint [ ]

SECTION 4

Overhauling engine units

USE THIS SPACE FOR LEARNER NOTES

Learning objectives

After studying this section you should be able to:

- Recognise techniques used to determine engine faults and wear.
- Understand removal and replacement procedures.
- Understand methods of finding faults.

Key terms

Blow-by   Gases escaping past the piston rings.
Piston crown   The top of the piston.
Clearance   A gap between two components.

www.plastigauge.co.uk

http://www.aa1car.com/library/engine2t.htm

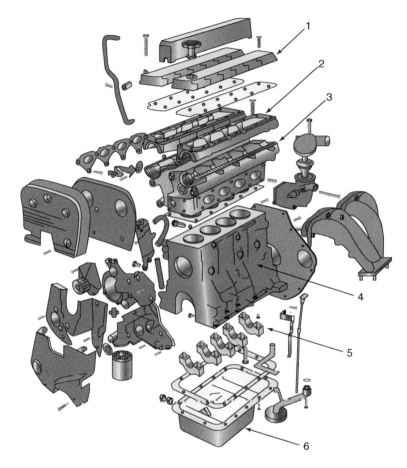

Engine exploded view

1. _____	4. _____
2. _____	5. _____
3. _____	6. _____

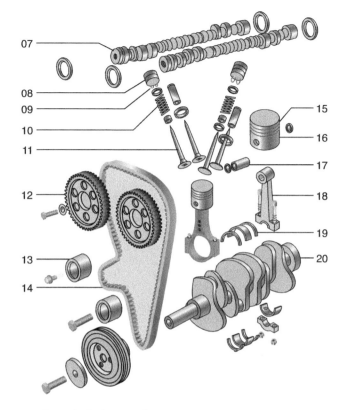

Name the engine parts shown in the images above.

07. _____	14, _____
08. _____	15. _____
09. _____	16. _____
10. _____	17. _____
11. _____	18. _____
12. _____	19. _____
13. _____	20. _____

 Removing, replacing and overhauling engines can involve heavy lifting so take care. It can also become complicated so observe manufacturer's instructions.

TIMING BELTS AND CHAINS

Timing belts, timing chains, cam belts do the same job. As the crankshaft turns the camshafts are also turned via the belt or chain.

What is important about correctly fitting a chain or belt?

_____

_____

What can happen if the timing belt or chain breaks?

_____

_____

List some easy ways to diagnose a snapped timing belt or chain.

● _____
  _____

● _____

● _____

Some engines are considered to be safe engines where the valves do not have enough reach to contact the pistons and cause damage.

A timing chain set up

When you remove a timing belt, what could suddenly happen to the camshaft and how can it be avoided?

_____

_____

_____

_____

When a timing chain or belt has been replaced what is a good procedure before starting the engine?

_____

_____

Timing belt

Add the following labels to the image of a timing belt above:

camshaft sprocket   **crankshaft sprocket**   **water-pump tensioner**   **idler pulley**

CAMSHAFTS

Today most manufacturers use overhead valve engines. When stripping down an overhead valve engine it would be common practice to remove the timing belt first and then inspect the camshaft, valve operating assemblies and hydraulic lifters.

When the camshaft has been cleaned, check the lobes for scoring, scuffing, fractures and signs of abnormal wear, also check the lubrication system and ensure galleries etc. are not blocked.

Name two good methods of diagnosing a worn camshaft lobe using measuring tools.

1 _____

2 _____

Worn hydraulic lifters

What is an important thing to do when removing the camshaft like that shown in the image opposite?

_____

_____

_____

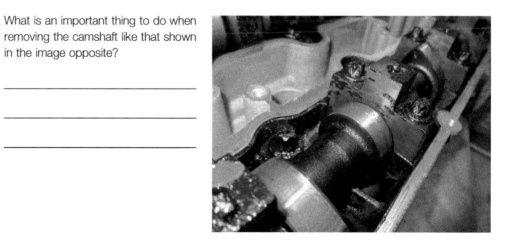

Camshaft and bearing caps

 **TIP** Camshafts look strong but they are in fact brittle and easily broken.

CYLINDER HEAD REMOVAL AND VALVE INSPECTION

On the diagram to the right number the cylinder head bolts to show the correct order of removal.

Once the cylinder head has been removed inspect the gasket for signs of wear. Check the cylinder head for cracks and damage.

The cylinder head should be checked to ensure it is flat and not warped, explain how this is done.

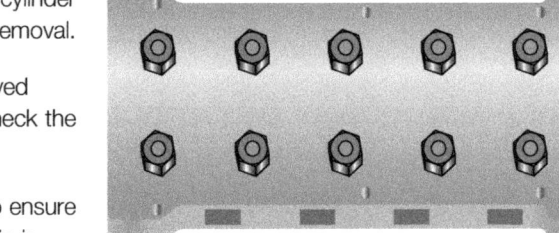

Head bolts

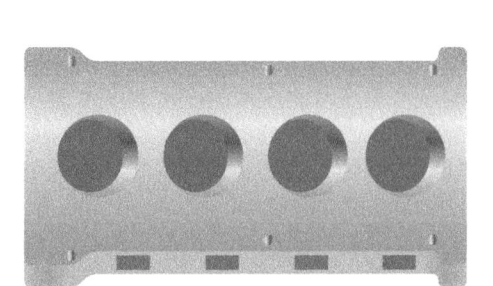

Checking for a level head

On the diagram above draw where measurements should be taken.

If the vehicle has been misfiring and valve problems are suspected, state a good diagnostic technique to check the valves are seating correctly.

TIP Valves should be removed using the correct valve spring compressor tool. Keep all parts in order.

Once removed, inspect the valves for condition and inspect the valve guides. In small groups, discuss the condition of the valve and guides shown in the images below.

Valves and guides

Once removed, how can you check if a valve is bent?

_____

If a valve is replaced or found to be leaking, then the valve seat will need to be matched to the valve, explain how this is achieved.

_____

_____

_____

_____

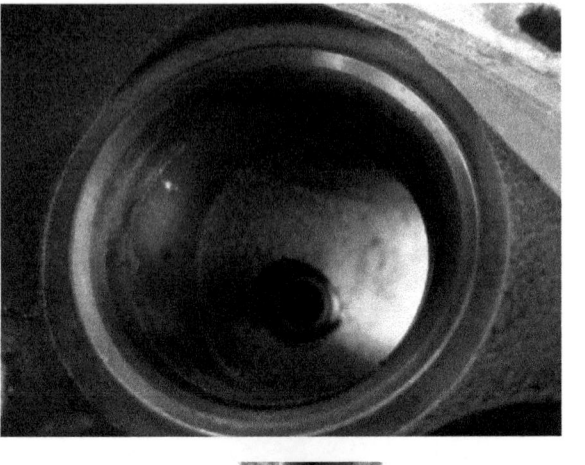

Valve and guide

VALVE STEM OIL SEALS

If the engine has been known to be burning oil it is worth replacing the valve stem oil seals whilst the cylinder head is removed, even though the issue may not be related to the oil seals. You should ensure correct installation and check with manufacturer's instructions to see if there are different seals for inlet and exhaust valves.

Label the diagram below which shows the installation of a typical valve stem oil seal.

 Always check with the manufacturer. Inlet and exhaust valves may use different seals.

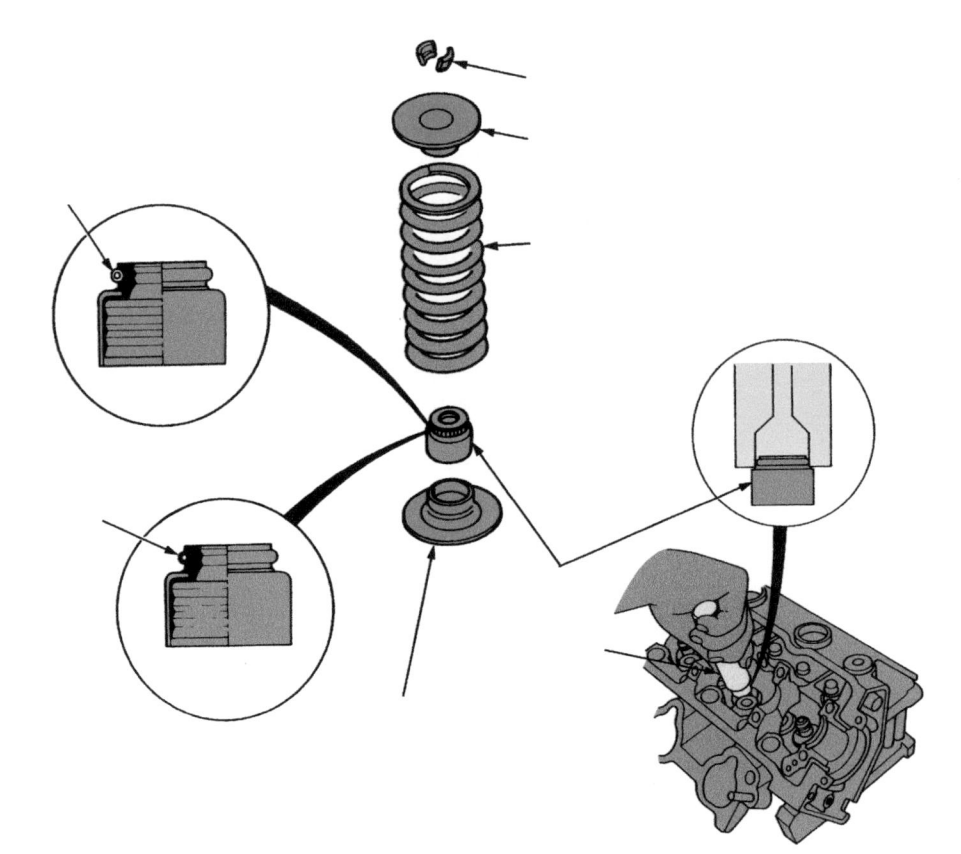

Valve stem oil seal

 Install the valve spring seats before installing the valve seals.

REPLACING HYDRAULIC LIFTERS

If hydraulic lifters are found to be faulty or are fitted with a new camshaft, what may be a good procedure before fitting?

_____

_____

_____

_____

PISTON PROTRUSION

How can piston protrusion be described?

_____

What is important about piston protrusion?

_____

_____

Which tools may be used to measure piston protrusion?

- _____
- _____
- _____

Explain how to measure piston protrusion using the tools you listed to answer the previous question.

- _____
- _____
- _____
- _____
- _____
- _____

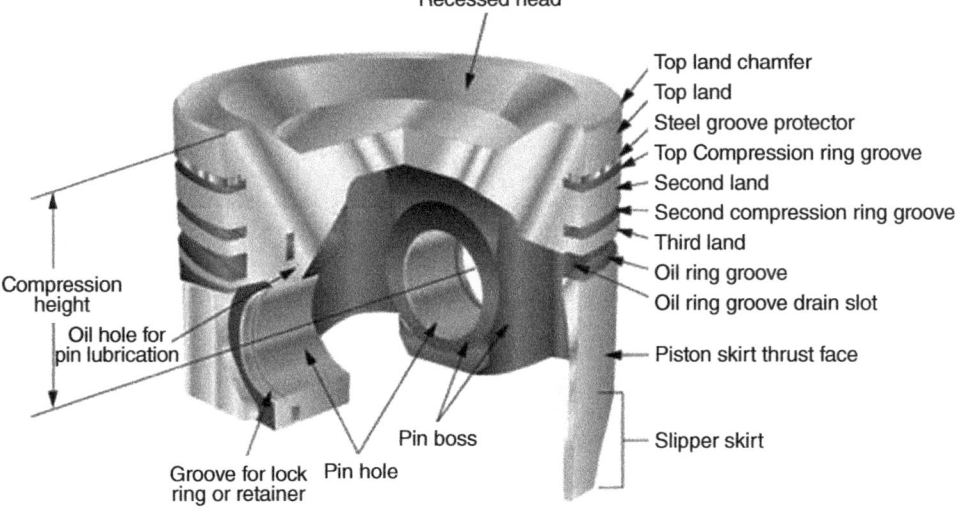

Measuring piston protrusion

BENT CONNECTING RODS

A bent connecting rod could be identified before removing the cylinder head using techniques already described in previous sections. What would identify a bent connecting rod?

_____

With the cylinder head removed it is often easy to confirm a bent connecting rod, but how?

_____

_____

_____

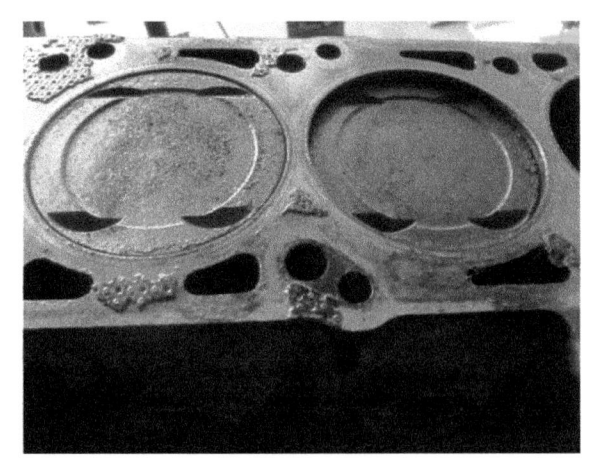

Bent connecting rods

PISTON RINGS

Piston rings serve three purposes, these are:

1 _____

2 _____

3 _____

CHECK Completing a cylinder leakage test or compression test may have helped you diagnose a damaged piston ring. Remove the piston, check and replace as required.

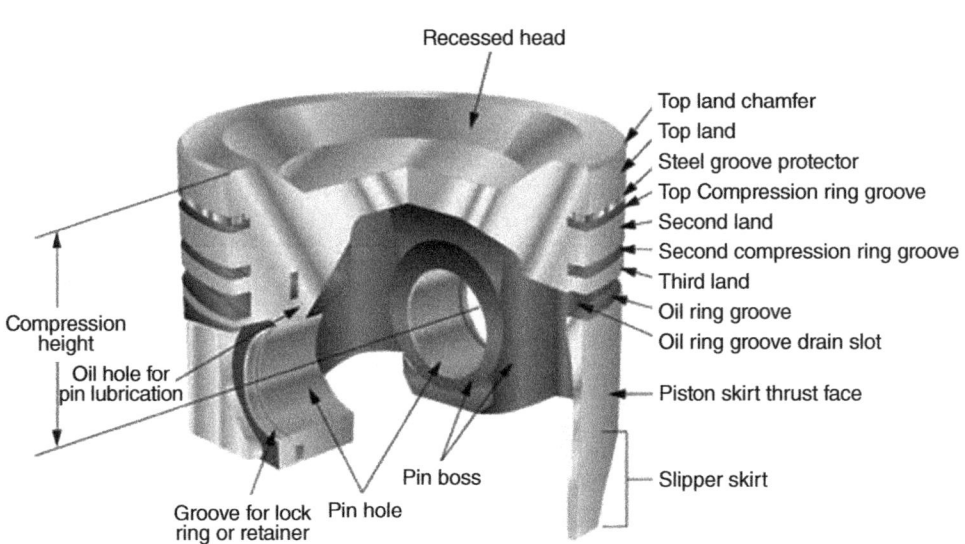

Recessed head

Top land chamfer
Top land
Steel groove protector
Top Compression ring groove
Second land
Second compression ring groove
Third land
Oil ring groove
Oil ring groove drain slot

Piston skirt thrust face

Slipper skirt

Compression height

Oil hole for pin lubrication

Pin boss

Groove for lock ring or retainer Pin hole

Piston ring assembly

CYLINDER BORE INSPECTION

The top of the cylinder bore is subject to high temperatures and pressures; also it is often not well lubricated. A cylinder must have the correct diameter, not be tapered or be out of round. The surface finish must also be in good condition to ensure the piston rings correctly seal and minimise **blow-by**.

What is the technician trying to measure in the diagram to the right?

_____

_____

_____

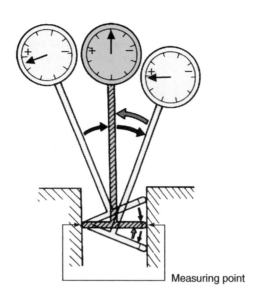

Measuring point

Measuring

CYLINDER BORE FINISH

The cylinder bore should be inspected for damage or wear. The desired finish should be coated in small criss-cross patterns, or a cross hatch affect, this is often achieved using a deglazing tool or honing tool.

Why would a criss-cross pattern be desirable?

_____

_____

 Research the process of honing a cylinder bore.

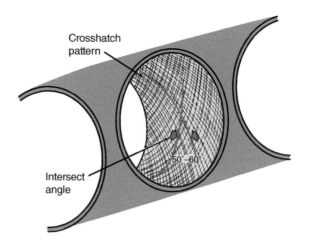

Cylinder bore after honing

CRANKSHAFT END FLOAT

Crankshaft end float can be measured using feeler blades but a more common and very accurate way is to use the method pictured below. Explain what is happening in the diagram below using the answer space provided on page 110.

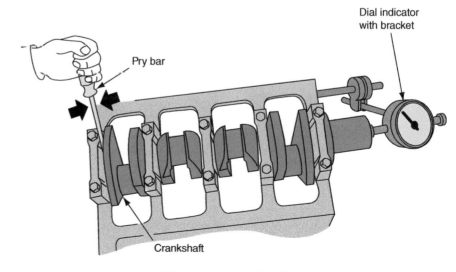

Measuring crankshaft end float

109

_____
_____
_____

CRANKSHAFT REMOVAL AND INSPECTION

The crankshaft main and big end bearings should be removed and inspected along with the crankshaft. If the crankshaft is showing wear or damage it can be re-surfaced by a professional. When fitting a crankshaft, bearing clearances are very important in relation to its operation, they must fit correctly.

Explain what might happen if the clearance is too tight.

_____
_____

What happens if the clearance is too large?

_____
_____

The bearing clearance can be measured in a variety of ways, you could use a micrometer and dial gauge or you can use a product called Plasigauge.

PLASTIGAUGE

Plastigauge is a simple but effective method of measuring the clearance between bearings and the crankshaft. It is particularly useful when feeler blades cannot easily be inserted for measuring. If necessary the measurement of the bearings can be taken without the removal of the crankshaft which can be easy and a quick method.

Visit this website **www.plastigauge.co.uk** and write some instructions on how to use this product using the pictures below.

Plastigauge

Write step-by-step instructions to measure big end bearings using Plastigauge below and on the following page.

- _____
- _____
- _____
_____
- _____
- _____
_____
- _____
_____

_____

_____

_____

_____

Scenario

Read the scenario and then complete the questions below.

A customer has arrived at the garage complaining of a misfire. On questioning the customer, he says he often checks the oil and coolant level, neither have dropped. A technician checks the ignition system and for a spark at the spark plugs, the spark plugs have lots of carbon on them and one is slightly damp. The technician dismisses this and decides to remove the cylinder head to check for further faults.

The technician removes the cylinder head and finds that one cylinder is clean as in the image below.

Clean cylinder

1 Describe what may have caused the clean cylinder.

_____

_____

_____

2 Describe how the technician could have diagnosed the job better.

_____

_____

_____

_____

3 Describe what else needs to be checked.

_____

_____

4 Discuss with your classmates the need to diagnose faults correctly and summarize the points covered using the space below.

_____

_____

_____

_____

_____

_____

_____

Multiple choice questions

Choose the correct answer from a), b) or c) and place a tick [✓] after your answer.

1   **A method to diagnose a worn camshaft would be?**

a) Using an engineer's rule [ ]

b) Using a Dial Test Indicator [ ]

c) Using feeler blades [ ]

2   **Crankshaft end float may be measured using:**

a) A Dial Test Indicator or feeler blades [ ]

b) A steel rule or feeler blades [ ]

c) An external micrometer or feeler blades [ ]

3   **A cylinder head can be checked for flatness using:**

a) A flat bar and feeler blades [ ]

b) A flat bar and internal micrometer [ ]

c) A flat bar and a flatness gauge [ ]

4   **Why is piston protrusion measurement important?**

a) It can affect the cooling of the cylinder [ ]

b) It can determine the thickness of head gasket to be fitted [ ]

c) It can affect the speed of the crankshaft [ ]

5   **Before stripping an engine, badly worn piston compression rings could be diagnosed using?**

a) A stethoscope [ ]

b) A camera inserted into the cylinder through the spark plug hole [ ]

c) A cylinder leakage test [ ]

SECTION 5

Hybrids and alternative fuels

USE THIS SPACE FOR LEARNER NOTES

_____

_____

_____

_____

_____

_____

_____

Learning objectives

After studying this section you should be able to:

● Recognise hybrid, electric and alternative technologies.
● Understand dangers of electric and hybrid vehicles.

Key terms

Hybrid   A vehicle that can operate using two sources of energy.
Generator   A device that can produce electricity.

www   http://hybridsmartcars.com/towing-hybrid-electric-car

http://www.nextgreencar.com/hybrid-cars

http://www.bmweducation.co.uk/cleanenergy/default.asp

http://world.honda.com/FuelCell

HYBRID VEHICLES

Due to the risk of death through electrocution you should only work on hybrid vehicles when you are trained and competent to do so. If you are in any doubt ask your supervisor.

What kind of vehicle could be classified as a hybrid?

_____

Why do most manufacturers use a combination of electric motors and a petrol engine rather than a diesel engine?

_____

_____

Manufacturers the world over have standardised the colour of high voltage wires in hybrid and electric vehicles, what colour is used?

_____

To help protect technicians from the dangers of electricity it is recommended that you wear insulated gloves; these are graded and will need to withstand the voltage being worked on.

List three checks to do on the gloves before use.

They must not be:

1 _____

2 _____

3 _____

High voltage gloves

TIP Trap some air in the glove to test for damage. Never use damaged gloves.

Some hybrid vehicles use voltages as high as 600 volts.

Hybrid vehicles use combined electric motors and generators. When you read information on hybrid vehicles the motor/generator may be referred to as 'M/G' and the term hybrid may be referred to as a 'HEV' or 'Hybrid electric vehicle'.

Motor/Generator

Research the difference between series and parallel hybrid vehicles, make some revision notes.

Most hybrid vehicles have a display for the driver to show if the battery is being charged using electric motors or the engine to power the vehicle.

Display

Some auxiliary electrical systems on hybrid vehicles may use high voltage, such as air conditioning, why is this?

_____

_____

Most hybrid vehicles still use a conventional 12v battery along with a high voltage battery, why is a 12v battery required?

_____

_____

High voltage batteries

 Hybrid batteries can be very powerful – over 200 volts.

Most manufacturers are generally using one of two types of batteries to power their vehicle, these are?

1 _____

2 _____

A problem with nickel is that the mining process is hazardous and it is considered to be carcinogenic.

CHECK Manufacturers provide a method to disconnect the battery to safely work on the vehicle; one example is shown in the image below which is a service plug which isolates the battery. Once removed you may have to wait for capacitors to discharge before working on the vehicle. Check manufacturer's data; if you are in doubt ask your supervisor and observe correct PPE.

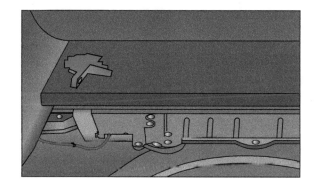

Service plug

The system should not allow the high voltage battery to completely discharge. Should this happen a special charger may be required which will only be available from the manufacturer.

The image opposite shows a vent from the battery storage area which you should advise customers never to cover up.

Why is there a need for this vent?

_____

_____

A high voltage battery is generally made up of many smaller cells.

Battery vent

Label the diagram below using the following terms:

HV battery module   **D-cell battery**
Single HV cell   **Single HV battery "stick"**

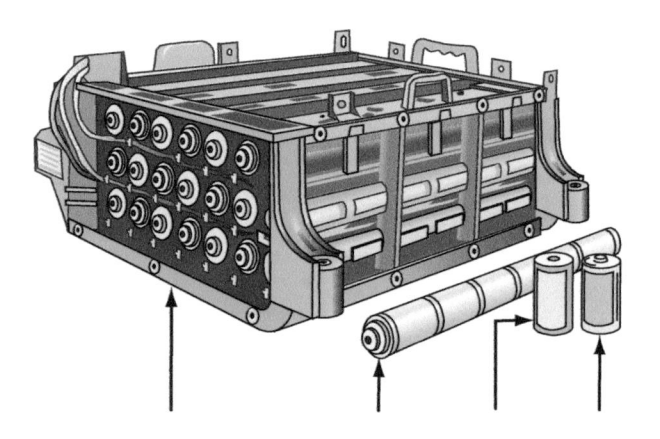

Hybrid vehicle battery

Invertor, converter and power boost assembly

The images below show a unit that acts as an invertor, converter and a booster. Explain the use of each using the space provided below and on page 117.

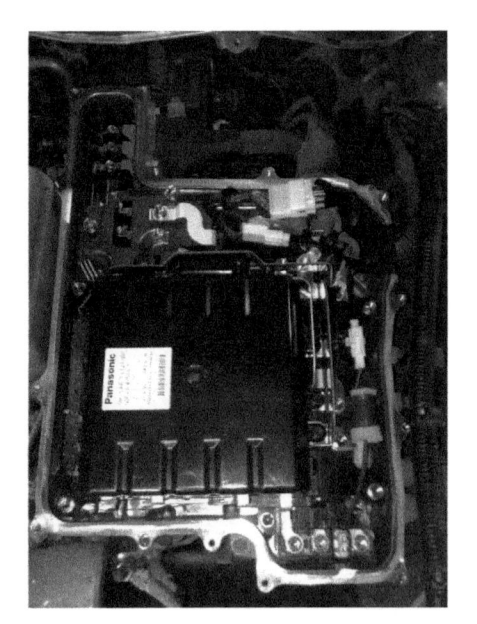

Invertor, converter and power boost assembly

Invertor

_____

_____

Converter

_____

_____

Booster

_____

_____

The system will also step down the voltage to charge the 12 volt auxiliary battery. Generally the units are checked with diagnostic testers and replaced completely rather than separate parts.

 The cover on the system shown below should not be removed unless you have had correct training.

Transaxle

Transaxle

What is meant by the term 'transaxle' in relation to hybrid vehicles?

_____

_____

As with a conventional gearbox, what should be checked during routine maintenance?

_____

What new check may need to be carried out during routine maintenance of a hybrid transaxle?

_____

_____

If the vehicle was to break down what should be considered before towing the vehicle?

_____

_____

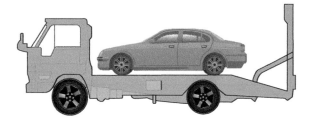 http://hybridsmartcars.com/towing-hybrid-electric-car

If in doubt, recommend that a flatbed truck is used to recover the vehicle so that none of the wheels arc turned.

Tow vehicle

Motor/generator

Any permanent magnet motor can be used as either a motor or generator. Motor power and electricity generation begin with the property of electromagnetism, what is meant by this term?

_____

_____

Explain the basic concept of a motor/generator.

_____

_____

_____

Why do manufacturers use AC motors and not DC motors?

_____

_____

When the vehicle is in motion the electricity powers the motors, when the vehicle is slowed down, the motor turns into a generator, but how does the generation of electricity slow the vehicle?

_____

_____

_____

 **CHECK** Many hybrid vehicles don't use conventional alternators or starter motors, the motor/generator completes these tasks and makes stop start function very well.

Hybrid car logo

EXTENDED RANGE/PLUG IN HYBRIDS

Manufacturers are now producing plug-in hybrid vehicles. What is the purpose of a plug-in hybrid car, and what happens when the main batteries become discharged?

_____

_____

_____

_____

_____

_____

_____

_____

_____

_____

Plug-in hybrid

Discuss in small groups the advantages and disadvantages of hybrid cars, and report to the rest of the class. State below their main advantages and disadvantages as compared to conventional cars.

Advantages	Disadvantages

Hybrid engines will start as soon as the battery voltage drops. This creates a dangerous situation when working on the engine, even draining oil. Make sure the system is isolated and remove the ignition key.

http://www.nextgreencar.com/hybrid-cars

Research different hybrid makes and models and find out about government incentives for owners.

Hybrid make	Hybrid model	Government incentives for owners

ELECTRIC VEHICLES

Electric vehicles, often referred to as Battery Electric Vehicles (BEV), have an average range of 100 miles. They do not have an engine as a backup if the battery becomes discharged through use. Drivers will need to consider their journey before setting off. List some situations that use the battery power quicker and would shorten the journey using the answer space on page 120.

- _____
- _____
- _____
- _____
- _____
- _____
- _____

A battery electric vehicle drives like any other car but it is quiet and requires charging from an external source when the battery discharges.

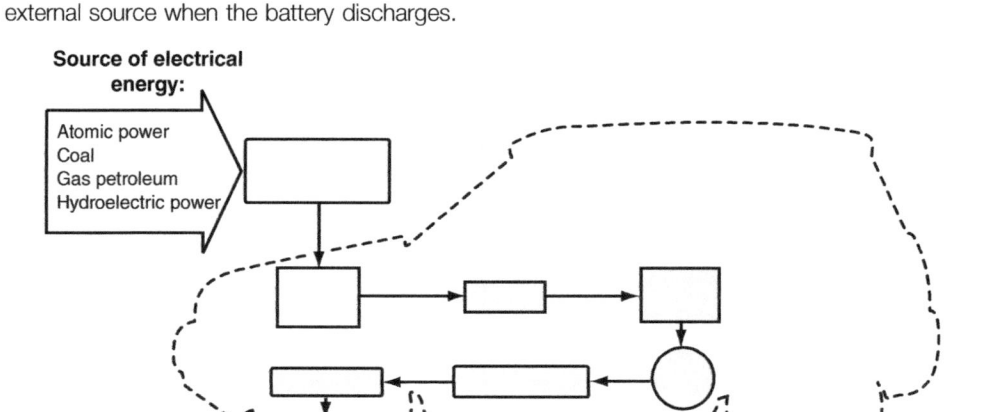

Source of electrical energy:

Atomic power
Coal
Gas petroleum
Hydroelectric power

Electric car layout

Add the following labels to the diagram of the 'Electric car layout' to demonstrate an electrical vehicle operation:

control device	**battery charger**	**battery**	**electrical power**
transmission	**motor**	**differential**	**supply**
wheel			

Although expensive, electric vehicles will have reduced road tax, will be exempt from congestion charges in inner cities and should have reduced maintenance costs. There is no oil and filter, air filter, fuel filter or spark plugs to replace.

Observe manufacturer's recommendations during repairs.

Although engine servicing is reduced electric vehicles will still require some maintenance.

List some items on an electric car that will require routine maintenance.

- _____
- _____
- _____
- _____
- _____
- _____
- _____

Electric vehicle

Many electric vehicles can simply be plugged into household mains sockets. This may take around eight hours to charge up. Some manufacturers may supply an extra socket for a fast charge which can be done at public places, however costs may be higher.

Electric vehicle charging

Electric vehicles may have many well designed features, such as solar panels in the rear windscreen to charge the 12 volt battery, electric motors housed in engine shaped containers so that customers feel comfortable and the battery range is considered to be electric fuel miles so that the gauge looks like a traditional fuel gauge. One well designed feature is that the batteries are in the lower part of the car below the seats, why is this feature important?

_____

HYDROGEN VEHICLES

 A video relating to hydrogen power has been produced by BMW and is available at the below link – although they suggest it's for younger students, it is highly recommended.

http://www.bmweducation.co.uk/cleanenergy/default.asp

Hydrogen is one of the most abundant elements on earth. Complete the following sentences regarding hydrogen.

Fossil fuels are combinations of _____ and _____; these are therefore known as

_____.

The combination of _____ and _____ forms water.

With the above statements in mind, consider where lots of hydrogen could be gathered to create an alternative to fossil fuels.

_____

How can hydrogen be produced from water?

_____

When would producing hydrogen NOT be environmentally friendly?

_____

_____

Why do we not see hydrogen used cars in mass production?

_____

_____

_____

 Research different forms of renewable energy and how they can be used.

FUEL CELL VEHICLES

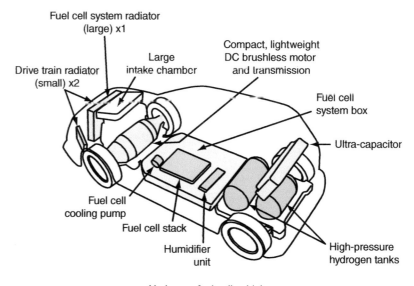

Fuel cell system radiator (large) x1

Large intake chamber

Compact, lightweight DC brushless motor and transmission

Drive train radiator (small) x2

Fuel cell system box

Ultra-capacitor

Fuel cell cooling pump

Fuel cell stack

Humidifier unit

High-pressure hydrogen tanks

Hydrogen fuel cell vehicle

A fuel cell is known as a Proton Exchange Membrane Fuel Cell (PEMFC). It is an electrical generation device. A chemical reaction takes place inside it, using hydrogen and oxygen to generate electricity to drive motors, this is basically the reverse of the principle of electrolysis. A limited amount of electricity can be produced so several cells are connected, this is known as a 'stack'.

Using hydrogen in this way does produce a 'by product' or 'exhaust'. What is this by product?

_____

Fuel cell vehicles also use an 'ultra-capacitor'. What is the job of a normal capacitor on a circuit board?

_____

Why might an 'ultra-capacitor' be used on a fuel cell vehicle?

_____

_____

List some of the difficulties that manufacturers will have to overcome for them to produce safe and reliable fuel cell vehicles.

- _____
  _____
- _____
- _____

 **http://world.honda.com/FuelCell/**

Research hybrid and electric vehicle qualifications. Colleges in your local area may offer extensive training in order to keep you safe, better understand the systems and assist with the diagnosis of faults.

Multiple choice questions

Choose the correct answer from a), b) or c) and place a tick [✓] after your answer.

1 **The colour of the high voltage wiring used in hybrid and electric vehicles is?**

   a) Red [ ]
   b) Blue [ ]
   c) Orange [ ]

2 **Li-ion batteries are called?**

   a) Lithuania – ion [ ]
   b) Lithium-ion [ ]
   c) Lithium iron [ ]

3 **Hybrid and electric vehicles use electric motors in their gearbox, for this reason extra care must be taken when?**

   a) Jacking up the vehicle [ ]
   b) Dealing with road side assistance and recovery [ ]
   c) Reversing the vehicle [ ]

4 **A hydrogen powered vehicle may incorporate a?**

   a) Fuel cell [ ]
   b) Fuel brace [ ]
   c) Fuel window [ ]

5 **A hydrogen powered car may incorporate an 'ultra-capacitor' what does this do?**

   a) Stores electric for sudden power boosts [ ]
   b) Stores electricity as a back-up for the main battery [ ]
   c) Stores hydrogen for boosting the engine [ ]

PART 4
CHASSIS SYSTEMS

USE THIS SPACE FOR LEARNER NOTES

_____

_____

_____

_____

_____

_____

_____

SECTION 1
Diagnosis and rectification of steering related faults 125

1   Introduction 126
2   Airbags 126
3   Power assisted steering (PAS) systems 128
4   Four wheel steering 136
5   Steering geometry 137
6   Multiple choice questions 143

SECTION 2
Diagnosis and rectification of suspension related faults 144

1   Suspension terms 145
2   The purpose of suspension 145
3   Passive and adaptive suspension 146
4   Electronic struts/dampers 148
5   Air suspension systems 149
6   Diagnosis of airbag suspension systems 152
7   Magneto-rheological dampers 152
8   Hydro-pneumatic suspension 154
9   Hydro-pneumatic suspension and braking systems 155
10 Multiple choice questions 157

SECTION 3
Diagnosis and rectification of braking related faults 158

1   Diagnosis and rectification of braking related faults 159
2   Brake fluid 160
3   Pads, discs and calipers 161
4   Roller brake tester 165
5   Calculating braking efficiencies 167
6   Electronically controlled handbrake systems 169
7   Anti-lock braking systems 170
8   Vehicle stability control (VSC) 175
9   Regenerative braking 176

10 Anti-lock brakes and associated systems diagnosis 177

11 Multiple choice questions 179

SECTION 4
Overhauling steering, braking and suspension units 180

1 Overhauling steering systems 181

2 Overhauling suspension systems 184

3 Overhauling braking systems 185

4 Multiple choice questions 189

SECTION 1

Diagnosis and rectification of steering related faults

USE THIS SPACE FOR LEARNER NOTES

Learning objectives

After studying this section you should be able to:

- Diagnose steering related faults.
- Understand methods of finding faults.
- Recognise how steering systems work, the parts involved and their operation.

Key terms

Torsion The action of twisting or turning.
SRS Safety restraint system, this may include airbags and seat belts.
PAS Power assisted steering, the system assists driver effort.

 WWW http://www.volkspage.net/technik/ssp/ssp/SSP_259.pdf

http://www.trw.com/sub_system/electrically_powered_hydraulic_steering

INTRODUCTION

As with many components, the problem with steering and suspension systems is that often parts are given two names but mean the same, such as a 'tie rod end' and 'track rod end' or 'bottom suspension arm' and 'wishbone' and then there is a 'hub' or a 'knuckle' and possibly many others. Throughout these sections try to learn a variety of names for the parts discussed so that you can recognise the terminology in tests or exams.

When diagnosing steering faults it is no good just diving in and replacing parts, you should find the fault that has caused the customer to complain, some other systems may be related to steering complaints and you should consider these. Correct diagnosis saves time and money and keeps customers happy.

 Ensure you wear correct PPE when completing steering work and dispose of waste correctly including metal components and power assisted steering fluids. Use the correct vehicle protection such as seat covers, floor mats and wing covers.

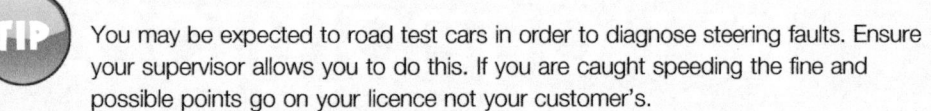 You may be expected to road test cars in order to diagnose steering faults. Ensure your supervisor allows you to do this. If you are caught speeding the fine and possible points go on your licence not your customer's.

Complete the table below to demonstrate how other systems can cause steering complaints. Add what fault the customer may complain of.

Fault	Customer complaint
Flat or low pressure front tyre	
Snapped engine auxiliary drive belt	
Bent suspension arm/wishbone	

Fault	Customer complaint
Sticking brake caliper piston	
Incorrect wheel alignment	
Wheel imbalance	
Incorrect steering geometry	
A warped brake disc	
Worn constant velocity (CV) joint	

AIRBAGS

The majority of vehicles on the road today have airbags installed including one in the steering wheel. An airbag may work in conjunction with seatbelt pre-tensioners and the system may be considered an SRS system.

What does SRS stand for? _____

The airbag is housed in the steering wheel and behind it there is an airbag squib (you may know this as a clock spring connector or rotary coupling).

Describe the purpose of the squib.

_____

_____

_____

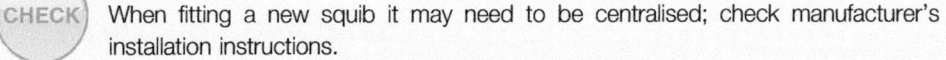 CHECK When fitting a new squib it may need to be centralised; check manufacturer's installation instructions.

Airbag squib

Label the diagram below of the airbag system using the terms listed opposite.

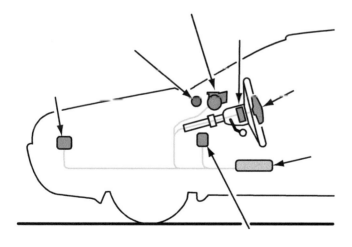

Airbag system

Add the following labels to the airbag system diagram:

forward sensor   **clockspring contact**   **in-car sensor**
warning light   **driver side airbag**
passenger side airbag   **electronic control module**

 NEVER check airbag systems with test lights, multi-meters or power probes whilst they are connected to a live circuit. ALWAYS follow manufacturer's instructions.

Before removing an airbag you must disconnect the battery, and then you must wait up to 10 minutes (depending on manufacturer) before working on the system. Why must you wait?

_____

_____

_____

The image to the left shows that the system has a forward crash sensor and an in car sensor often called a 'safing', 'arming' or 'safety' sensor, what is the purpose of having two sensors?

_____

_____

Label the squib coil and igniter charge on the airbag module diagram below.

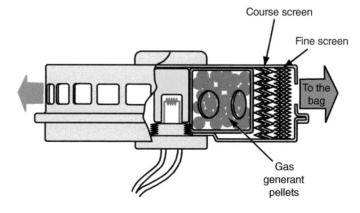

Airbag module releasing nitrogen gas

127

Write a step-by-step plan of how to remove an airbag.

The following is a typical sequence of events for airbag deployment in a frontal collision:

1 **Time zero – impact begins and the airbag system is idle.**
2 **20 milliseconds – The sensors are sending an impact signal to the airbag module.**
3 **23 milliseconds – The airbag is inflated and is up against the occupant's chest. The occupant's body has not yet begun to move as a result of the impact.**
4 **40 milliseconds – The airbag is almost fully deployed and the occupant's body begins to move forward because of the impact.**
5 **70 milliseconds – The airbag begins to absorb the forward movement of the occupant's body.**

Complete the sentence below using the words from the list.

nitrogen	**chemical**	**reaction**	**sensors**
airbag	**sodium**	**igniter**	

When the crash _____ agree an accident has taken place, the system inflates the

_____; an electrical current is sent from the electronic control unit to an

'_____', this starts a _____ reaction and a gas (usually _____)

fills the airbag in a fraction of a second. The screens may filter _____ hydroxide dust

from the gas which is created during the chemical _____.

The only customer complaint regarding an airbag should be that the warning lamp is illuminated. You must always follow manufacturer's instructions when diagnosing faults. A code reader is normally required for correct diagnosis along with visual inspections of connectors and wiring. Modern vehicles may have many airbags fitted; a fault with any of them will illuminate the warning lamp. Describe how the system completes a self-test using the warning lamp.

_____

_____

_____

_____

_____

Research how seat belt pre-tensioners operate, find at least two different methods.

POWER ASSISTED STEERING (PAS) SYSTEMS

Hydraulic systems

The first part of the hydraulic power steering system that needs regular checks is the drive belt. List some faults that would require a new belt to be fitted.

- _____
- _____
- _____
- _____
- _____

The belt should be adjusted to the correct tension; this may involve checking it with a belt tension gauge.

Belt tension gauge

Engine driven power steering pump

The power steering pump is turned by the crankshaft via a belt. Most manufacturers use a rotary vane pump. Some belts have self-adjusting tensioners to keep the perfect tension and help prevent slip when the belt wears or stretches. Can you name the self-tensioner on the below left?

As this pump turns with engine speed, what happens when engine revs are high?

_____

_____

 Research how a rotary vane power steering pump works, draw a sketch with your explanation.

Rack and pinion system

Name the numbered parts on the steering rack diagram on page 130:

1 _____

2 _____

3 _____

4 Fluid inlet from the pump (high pressure)

5 Fluid return to the pump (low pressure)

6 _____

7 _____

8 _____

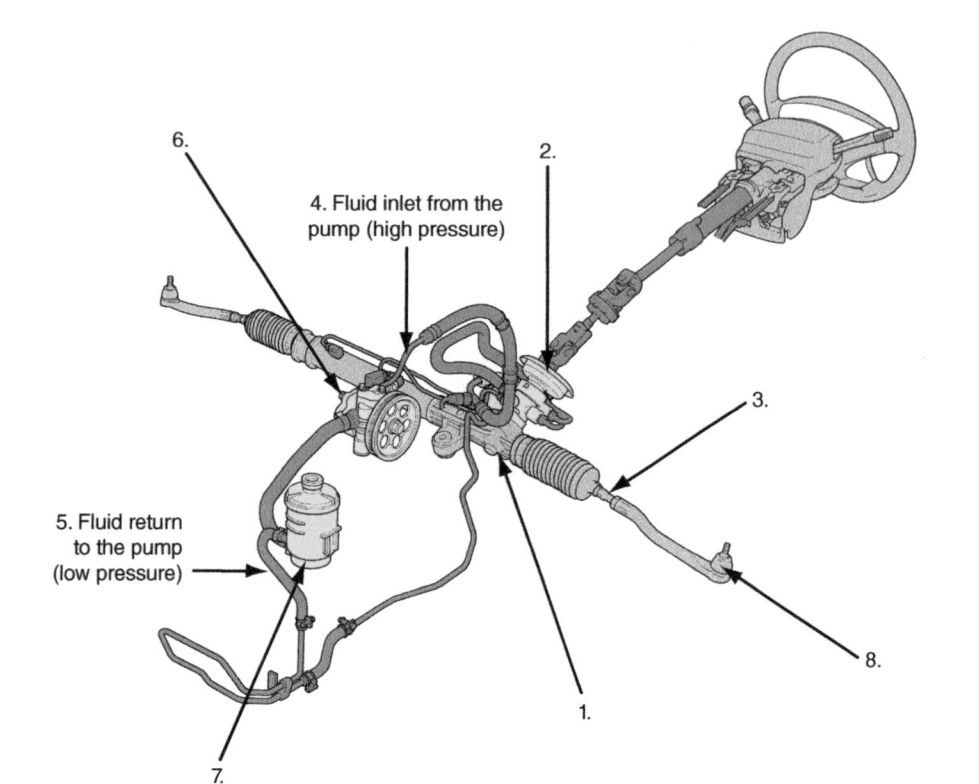

6.

4. Fluid inlet from the pump (high pressure)

2.

3.

5. Fluid return to the pump (low pressure)

8.

1.

7.

Steering rack layout

The pinion incorporates a **torsion** bar which allows the power steering fluid to be directed to one side of the rack or the other, providing assistance to the driver. At straight ahead position a small amount of pressure is generated and fluid flows from the inlet, through the valve to the return. Describe what happens when the steering is turned.

_____

_____

_____

_____

Steering rack spool

Steering gearbox system

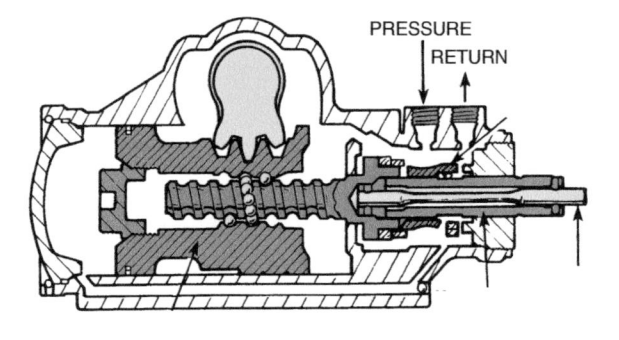

PRESSURE
RETURN

Steering gearbox

Add the labels to the above diagram:

spool valve   **torsion bar**   **input shaft**   **piston**

A conventional steering recirculating ball steering box can also utilise power steering; the fluid is directed through the spool valve which is turned by the torsion bar and acts on both sides of the piston.

Research to find out why the ball bearings are fitted to the re-circulating ball steering box shown in the diagram on page 130.

Diagnosis

As well as checking for free play and visually checking the system for leaks, completing a hydraulic pressure test on the power steering system can be helpful in diagnosing faults.

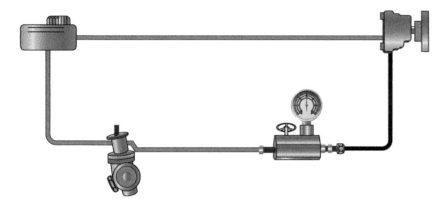

Checking PAS pump pressure

When completing a pressure test ensure the equipment is correctly fitted following manufacturer's instructions; label the diagram above and explain where the gauge should be fitted in the system.

_____

_____

To fit the gauge you will be required to disconnect a pipe on the hydraulic system. What needs to be completed now and how is it done?

_____

_____

_____

 CHECK There are various tests that can be completed to diagnose faults. Care must be taken to follow vehicle manufacturer's instructions and the tool operating instructions as closing the valve for too long can cause damage to the system.

Complete the table below by stating the possible fault and how it should be rectified.

Pressure reading from the pump to the gauge is low.	_____ _____ _____ _____
There is an excessive pressure difference when tested at 1000 RPM and 3000 RPM	_____ _____ _____
A low pressure reading when on full lock	_____ _____ _____ _____

Electro-hydraulic power steering system (EHPS)

Electro-hydraulic power steering is a similar system to a conventional system but rather than a mechanically powered pump, an electric motor is used. What are the main advantages of this system?

- _____

  _____

- _____

  _____

  _____

The system shown below incorporates a steering angle sensor and an electronic control unit. What is the purpose of these parts?

_____

_____

_____

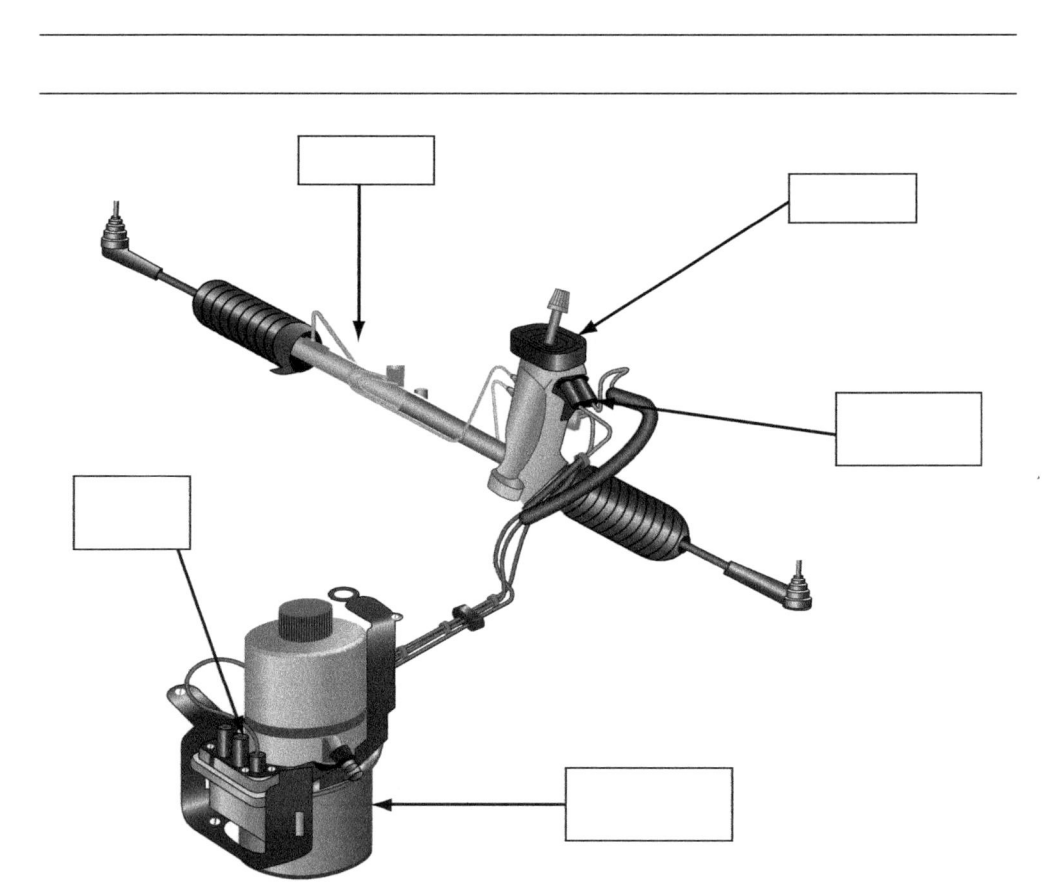

Steering system

Add the labels to the diagram above:

steering angle sensor   **steering gear pump**   **electronic control unit**
steering rack   **with motor**   **spool valve**

The diagram below shows a simplified drawing of the steering angle hall sensor.

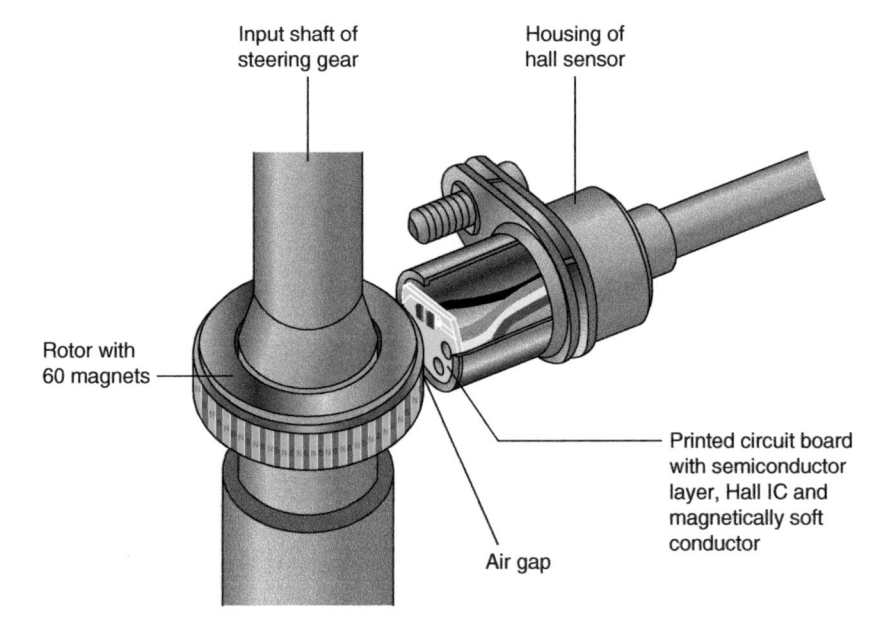

Input shaft of steering gear

Housing of hall sensor

Rotor with 60 magnets

Printed circuit board with semiconductor layer, Hall IC and magnetically soft conductor

Air gap

Steering angle sensor

 Research electro-hydraulic power steering. Discuss three checks to complete on the system.

 **http://www.volkspage.net/technik/ssp/ssp/SSP_259.pdf**

Describe how the steering angle sensor works.

_____

_____

_____

_____

Electro-hydraulic steering rack

Explain how faults on this system may be diagnosed.

_____

_____

_____

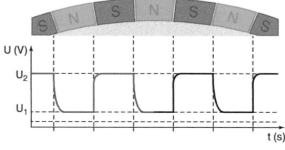

WWW http://www.trw.com/sub_system/electrically_powered_hydraulic_steering

Should a serious fault occur with the electronic system it should 'fail safe'. What is meant by this term?

_____

_____

_____

Hall sensor operation

If the magnet is in direct contact with the sensor, this position is known as a magnetic barrier, a Hall voltage is generated in this state. The magnitude of the voltage depends on the intensity of the magnetic fields between the magnets. As soon as the relevant magnet rotates further round and leaves the magnetic barrier, the magnetic field is deflected by the Hall sensor integrated circuit, the voltage drops in the circuit and the Hall sensor switches off.

The 'U' in the diagram below basically refers to the sensor output signal; this may be referred to as 'hall voltage offset'. It is simply voltage and as you can see is measured over time.

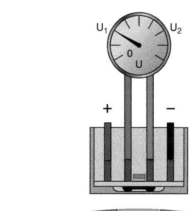

Hall sensor

TIP If a customer complains of a 'heavy steering' fault then don't forget to check the basics, including tyre pressures.

What does this warning symbol mean?

_____

_____

_____

_____

Fully electric power steering

Some vehicles have limited space under the bonnet so some manufacturers have developed electric power steering where the motor is attached to the column.

Electric power steering saves under bonnet space and makes the engine more fuel efficient. List some other advantages of electric power steering.

_____

_____

_____

_____

The system requires the use of an electronic control unit which contains 'control algorithms' to provide good performance of the steering system. Describe what is meant by the term 'control algorithms'.

Electric power steering

_____

_____

The system needs the following parts to operate:

- **A powerful electric motor which is geared to the steering column**
- **An electronic control unit**
- **Road speed sensors**
- **Steering torque sensors**

What is the purpose of the road speed sensor?

_____

_____

What is the purpose of the steering torque sensors?

_____

_____

Steering angle sensor

It's important to realise that this measured torque is a two-way process – if the front wheels are on wet grass what affect will it have on the steering?

_____

_____

_____

When the vehicle is on dry tarmac what will be the effect on the steering?

_____

_____

Electric motor and ECU

The motor used is a DC motor, often attached to the column or directly to the steering rack, how is the direction of rotation changed to turn the steering one way or the other?

_____

Research vehicle parts websites to ensure you can identify electric power steering parts.

Steering task

Enter your place of training or workplace and locate a vehicle with either electric or hydraulic power steering.

Imagine you have diagnosed a fault on the steering rack on that vehicle; the customer has given authorisation to have it replaced.

Make a list of the tools and equipment needed to complete this task.

- _____
- _____
- _____
- _____
- _____
- _____
- _____
- _____
- _____

List all of the parts you will need to complete this task, this may include hose clips, split pins, oil etc.

- _____
- _____
- _____
- _____

Locate a manual or some data sheets and list any technical information needed to complete this task. This may include oil quantities, torque settings etc.

- _____
- _____
- _____

Describe how you will dispose of waste products.

- _____
- _____

Write a brief description of how to complete the task.

_____

_____

_____

_____

_____

FOUR WHEEL STEERING

There are only a few manufacturers in the world that have produced cars with four wheel steering systems. Modern systems are generally controlled by an ECU using information from turn angle and vehicle speed sensors.

The steering of the rear wheels in relation to the front wheels may depend on the speed of the vehicle.

On the diagrams below, add the rear set of wheels to show their direction.

Low speed   High speed

4 wheel steering

Explain what happens at low speed and the main advantage.

_____

_____

What happens as the vehicle starts to speed up?

_____

_____

_____

The rear wheels only turn a small amount in comparison to the front wheels, especially at high speed.

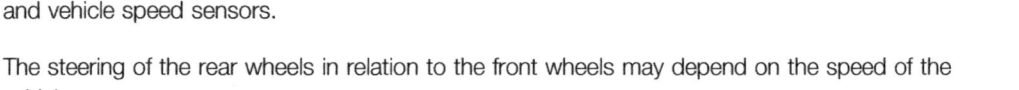

Research the types of vehicles that have four wheel steering systems and investigate further how the system works including advantages and disadvantages.

STEERING GEOMETRY

Steering geometry extends further than just tracking or wheel alignment. Manufacturers design cars to hold the road perfectly and to ensure that the tyres wear correctly and evenly, there is a link with suspension parts when we discuss steering geometry.

Caster angle

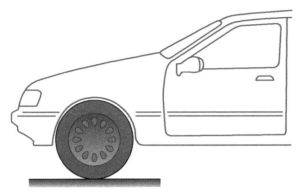

Caster angle

On the diagram above, mark a centre line and a line to demonstrate positive caster and a line to demonstrate negative caster.

How can you explain caster angle?

_____

_____

_____

What is the purpose of caster angle?

_____

_____

_____

What may affect the caster angle?

_____

_____

_____

_____

If castor angle was out of specification, what may the driver notice?

_____

_____

_____

CHECK Measuring caster angle may depend on the equipment used so always follow manufacturer and tool maker instructions.

Explain a popular method of checking caster angle.

_____

_____

_____

_____

Camber angle

Add lines to the diagrams below to show positive and negative camber angle.

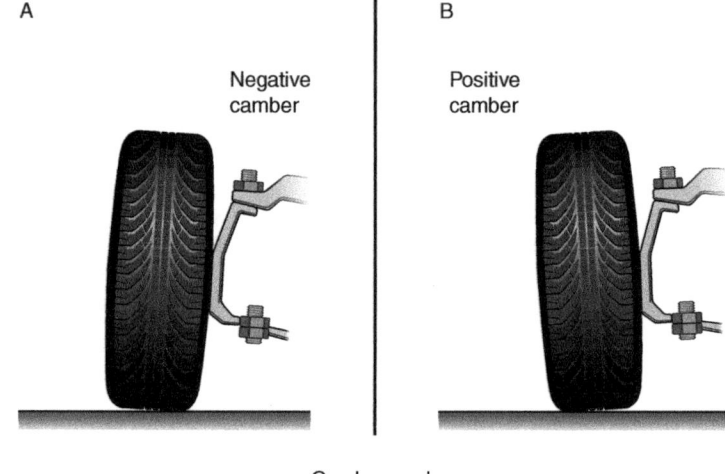

A

Negative
camber

B

Positive
camber

Camber angle

In the space above draw a diagram to demonstrate camber angle based on a Macpherson strut type suspension system. Include positive and negative camber.

Explain camber angle.

_____

_____

_____

If the camber was set incorrectly what would you notice when you inspect the vehicle?

_____

_____

_____

> Investigate different methods of adjusting caster and camber on different types of suspension designs, find three websites, compare the information and share them with your classmates.

Steering axis inclination (SAI)

Steering axis inclination is viewed from the front, it is an engineering angle designed to project the weight of the vehicle to the road surface for stability. The angle is measured with a line through the centre of the strut, or upper and lower ball joints depending on design and true vertical through the centre of the wheel. Add lines to the diagram opposite to demonstrate SAI.

Steering axis
inclination

Scrub radius

You will notice that the diagram below shows steering axis inclination and the lines do not meet at the road surface. This causes a 'scrub radius'.

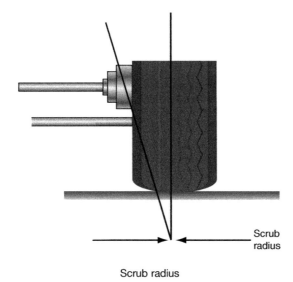

Scrub radius

Scrub radius is designed by the manufacturer and is set at the factory when the vehicle is constructed; however, scrub radius can still be affected. What would affect scrub radius?

● _____

● _____

As the vehicle corners at high speed how can this change scrub radius?

● _____

_____

 Research scrub radius and wheel offset. Draw some diagrams to explain them and how they are linked.

Complete the table below to demonstrate the difference in intersection of the lines.

Intersection of the lines of scrub radius	Meaning
Lines intersect below the road surface	
Lines intersect at the road surface	
Lines intersect above the road surface	

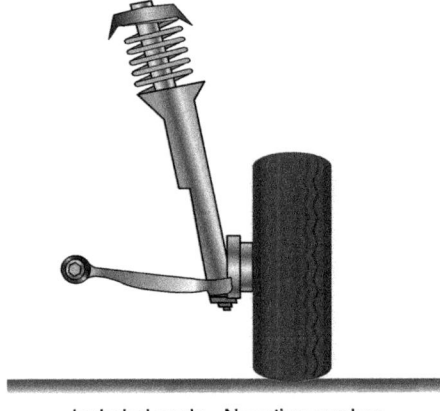

Included angle - Negative camber

Excessive scrub radius, positive or negative, can cause increased steering effort and road feel increases as the steering becomes susceptible to road shocks.

What is a common way of changing scrub radius without it being realised?

_____

_____

_____

Included angle

Steering axis inclination has another relationship with a steering angle; known as the included angle.

The included angle is steering axis inclination and camber angle: if the camber is positive, then this is added to the steering axis inclination to create an included angle, if the camber is negative then the angle is taken from the steering axis inclination to create the included angle.

Add lines to the diagrams on the right to show steering axis inclination and positive and negative camber to create an included angle.

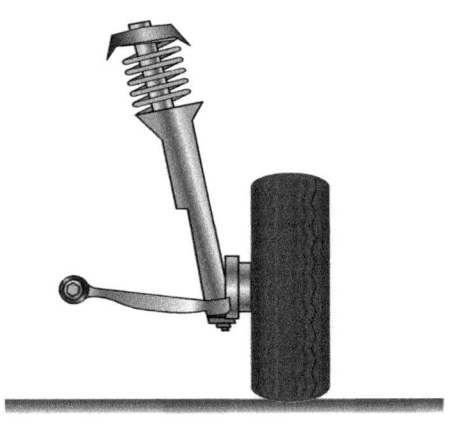

Included angle - Positive camber

 **TIP** Comparing camber, steering axis inclination and included angle can help to identify worn or damaged components.

Toe out on turns

Toe out on turns may also be referred to as Ackerman angle.

Show toe out on turns on the sketch opposite and explain what it means.

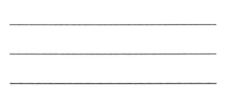

_____

_____

_____

_____

_____

Describe how to check toe out on turns.

_____

_____

_____

_____

_____

_____

Toe out on turns

 Enter your place of training or study and measure toe out on turns.

Thrust angle

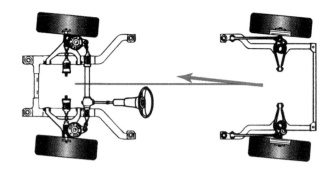

Thrust angle

The figure above shows a geometric centre line through the centre of the vehicle and a red arrow to show the thrust angle of the rear axle.

Complete the sentence below in regards to thrust angle, add the following words:

straight	**toe**	**front**
rear	**thrust**	

If the rear _____ is out of specification then the vehicle will tend to run in the direction of

the _____ angle rather than _____ ahead. This is why when the vehicle design

allows you should adjust the _____ tracking before you adjust the _____ tracking.

Steering geometry preliminary checks

Before checking a vehicle for alignment you should first carry out some preliminary checks to the vehicle. List below and on page 142 some checks to be completed.

- _____

  _____

- _____

- _____

- _____
- _____
- _____
- _____
- _____

 **www** http://www.motor.org.uk/documentlibrary/Sep%2009/TT%20_%20Sept%2009.pdf

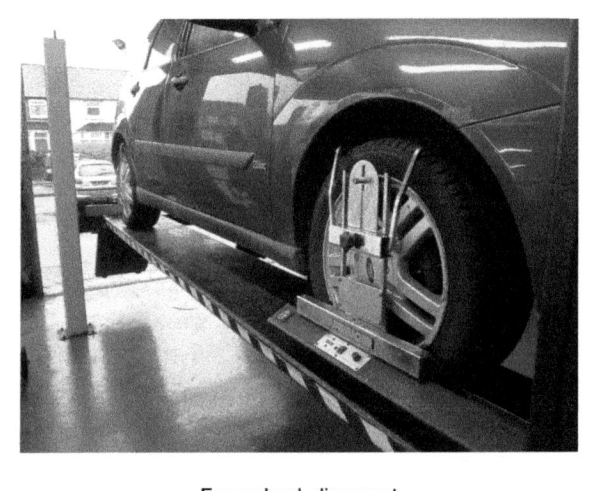

Four wheel alignment

Research different tools and equipment used in wheel alignment checking and adjusting. List advantages and disadvantages of different equipment.

Advantages

Disadvantages

 Scenario

A customer complains that their power steering appears difficult to turn.

Suggest two questions to ask the customer.

1 _____

2 _____

_____

You check the vehicle and notice it has electric power steering, what should be the next step in your diagnosis?

_____

_____

_____

_____

_____

_____

_____

Which non-steering related systems would it be useful to check?

_____

_____

_____

If you are required to replace major steering components, what should you do before returning the vehicle to the customer?

_____

_____

_____

After completion of the job the car still pulls to the left very slightly, what should you do now?

_____

_____

_____

_____

Multiple choice questions

Choose the correct answer from a), b) or c) and place a tick [✓] after your answer.

1 **Which one of the following components provides an input to the ECU to control an electric power steering system?**

   a) Stepper motor [ ]

   b) Electric motor [ ]

   c) Torque sensor [ ]

2 **When carrying out electrical repairs to a vehicle's steering system, which one of the following sources of information, will provide appropriate instructions?**

   a) Service record book [ ]

   b) Workshop manual [ ]

   c) Owner's handbook [ ]

3 **Why is it important to record all workshop activities?**

   a) It is a legal requirement [ ]

   b) They can be accessed in case of a dispute [ ]

   c) Warranties will be cancelled if they are not available [ ]

   d) Managers will use them to assess the company financial state [ ]

4 **A vehicle has failed its annual DVSA test on a split steering rack gaiter. The correct procedure will normally be?**

   a) Remove and replace the entire steering rack [ ]

   b) Remove and replace the steering column [ ]

   c) Remove and replace just the gaiter [ ]

5 **During a routine vehicle service what should be done to an EHPS system?**

   a) Check the oil level [ ]

   b) Adjust the drive belt [ ]

   c) Complete a vehicle alignment [ ]

6 **How long does it take for an airbag to fully deploy?**

   a) Less than 1 second [ ]

   b) 1 – 2 seconds [ ]

   c) 2 – 3 seconds [ ]

SECTION 2

Diagnosis and rectification of suspension related faults

USE THIS SPACE FOR LEARNER NOTES

_____

_____

_____

Learning objectives

After studying this section you should be able to:

● Diagnose suspension related faults.
● Understand methods of finding faults.
● Recognise how suspension systems work, parts involved and their operation.

Key terms

Damper A shock absorber.
Adaptive suspension Suspension that can generally be changed during operation by an electronic control unit.
Passive suspension A suspension system which works as designed and cannot be adjusted for different driving styles.

http://www.bosch.co.uk/en/uk/startpage_2/country-landingpage.php

http://www.rangerovers.net/repairdetails/airsuspension/index.html

http://www.citroenet.org.uk/miscellaneous/hydraulics/hydraulics-1.html

http://www.carbibles.com/suspension_bible.html

SUSPENSION TERMS

Research the following terms and explain their meaning.

Bump _____

_____

Rebound – _____

_____

Shimmy – _____

_____

_____

Sprung weight – _____

_____

Un-sprung weight – _____

_____

Semi-elliptical spring – _____

_____

Ride height – _____

_____

_____

Compliance bushes – _____

_____

Suspension system

THE PURPOSE OF SUSPENSION

Discuss with your classmates/colleagues what you should expect from a good suspension system. Make a list of possible complaints:

- _____
- _____
- _____
- _____
- _____
- _____

What is the purpose of a suspension spring?

_____

What is the purpose of a suspension damper?

_____

_____

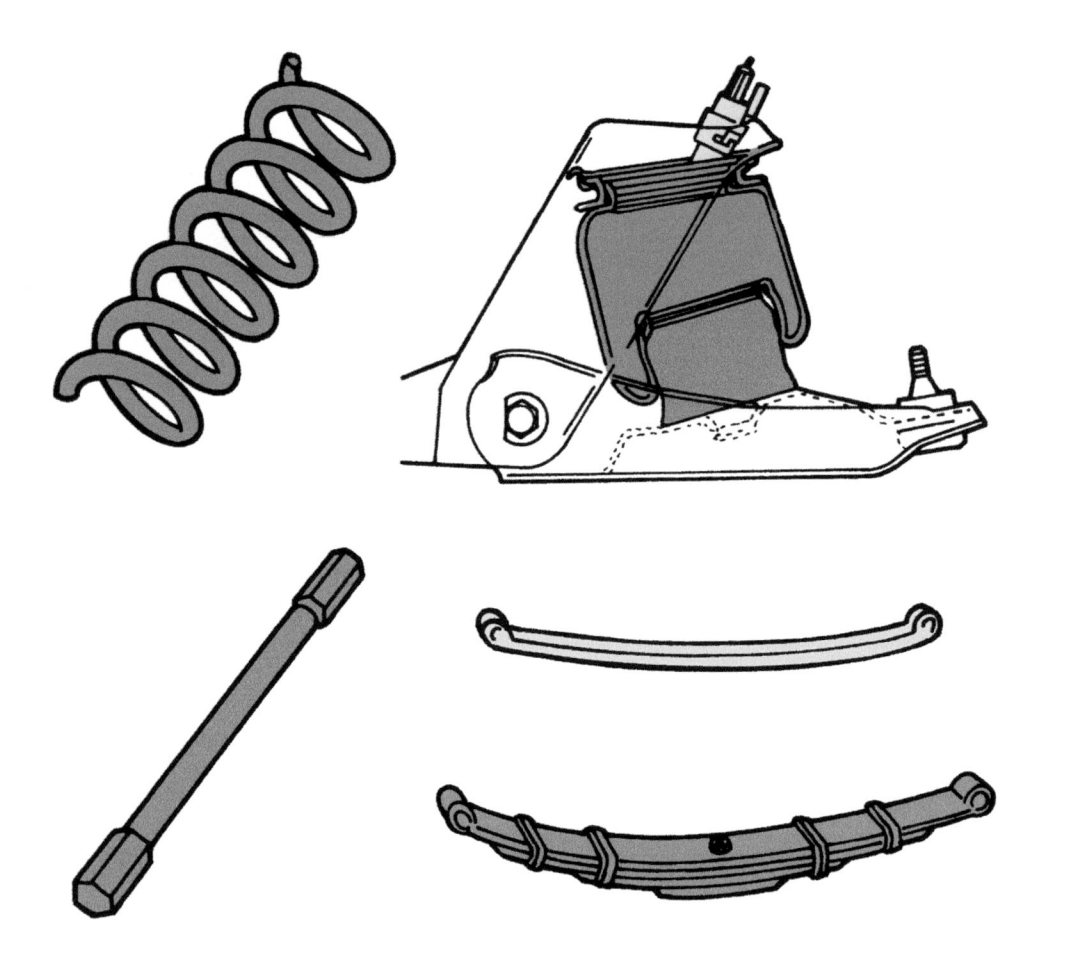

Suspension parts

The image above shows different suspension springs, name them.

PASSIVE AND ADAPTIVE SUSPENSION

Passive and adaptive (sometimes called active) are common terms when it comes to suspension systems but what do they mean?

Passive suspension – _____

_____

_____

_____

Passive suspension system

Adaptive suspension – _____

_____

Adaptive suspension

How does adaptive suspension differ from passive suspension?

_____

_____

_____

Sensors

Sensors used in adaptive suspension can be similar to those used in other parts of the vehicle. For example, steering angle sensors are used with electric power steering, throttle position sensors are used with engine management systems.

Yaw rate sensor

WWW http://www.bosch.co.uk/en/uk/startpage_2/country-landingpage.php

What does a yaw rate sensor measure?

_____

_____

Where is the yaw rate sensor often situated in the vehicle?

_____

_____

How does a yaw rate sensor work? Continue your answer in the space provided on page 148.

_____

_____

_____
_____
_____
_____
_____

Self-levelling sensors

Self-levelling of suspension can be important to some vehicles; towing or carrying heavy loads can affect tyre wear and vehicle handling. When a load is carried, headlamps can also be raised in the air which dazzles othe drivers. Self-levelling suspension can counter act these problems.

Commonly used with vehicles using air suspension, a self levelling sensor will be connecetd to which components?

_____
_____
_____

CHECK It is common for manufacturers to fit rotary Hall effect sensors as suspension height sensors.

 **WWW** http://www.rangerovers.net/repairdetails/airsuspension/index.html

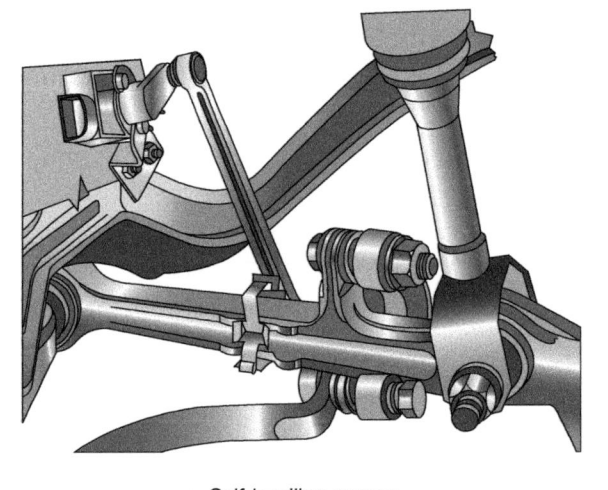

Self-levelling sensor

 Research

Research suspension components and sensors so that you can recognise them on vehicles.

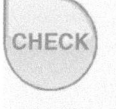 **CHECK** Some self-levelling units may need to be calibrated if they are replaced or when suspension parts are replaced, this will normally require the use of a code reader. Check with the manufacturer how this may be done.

ELECTRONIC STRUTS/DAMPERS

The basic idea behind simple electronically controlled suspension is to use electronically adjustable dampers and/or struts so suspension ride control characteristics can be adjusted or adapted to changing driving conditions, resulting in improved ride and handling.

How can a basic hydraulic damper or strut vary the dampening effects on the spring?

_____

_____

_____

Add arrows to the diagram below to indicate where fluid movement occurs in the damper.

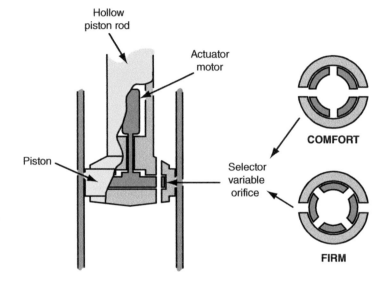

Modern dampers and struts may use magnetic valves; these can work very quickly and allow an ECU to quickly adjust the system which can even prevent the vehicle nose from dipping under harsh braking.

How do modern electronic dampers and struts work?

_____

_____

_____

_____

AIR SUSPENSION SYSTEMS

What is the main difference between air suspension and conventional suspension components?

_____

Some air suspension systems can adjust the suspension in real time as the vehicle hits irregularities in the road surface, allowing a smooth and comfortable ride.

List below the advantages of air suspension:

- _____
- _____
- _____
- _____
- _____

A typical suspension system will require a number of parts to operate the system, can you list the parts required?

- _____
- _____
- _____
- _____
- _____
- _____

Label the air suspension parts on the diagram below.

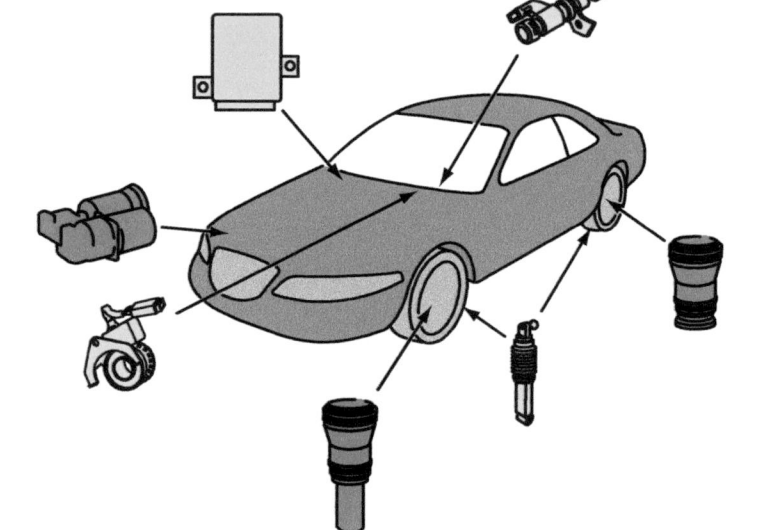

Typical air suspension system

The air suspension control module

What is the purpose of the air suspension control module?

_____

_____

_____

_____

Typical air suspension control module

TIP If replaced the module will probably need programming to the vehicle and may need calibrating after repairs.

Air springs

Air suspension springs may incorporate a damper or may be separate to the damper.

What is the purpose of the aluminium sleeve around the airbag?

_____

_____

_____

Air suspension spring

Air suspension height sensors

Which kind of height sensors are commonly used in air suspension systems?

_____

_____

Air suspension height adjuster

Air suspension compressor

The air suspension compressor obviously pressurises airbags and the reservoir but it also has an exhaust. What is the purpose of the exhaust?

_____

_____

_____

_____

_____

_____

Air suspension compressor

Air suspension reservior

Why might the system need a reservoir?

_____

_____

_____

_____

_____

Air suspension reservoir

Receiver drier

When air is compressed condensation is generated, moisture can create problems with freezing and corrosion so a receiver drier is used.

How does a receiver drier work?

_____

_____

_____

_____

_____

Receiver drier

Air reservoir valve block

The air suspension system uses valve blocks, what do the valve blocks do?

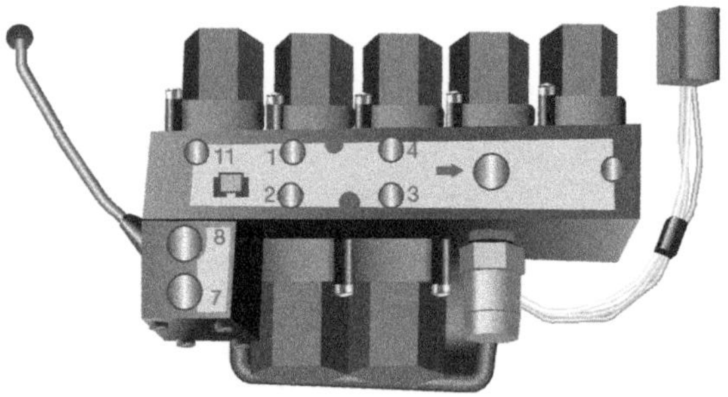

_____

_____

_____

_____

_____

Valve blocks

⚡ Airbag pressures may vary dependent on manufacturer and could be quite high, ALWAYS follow manufacturer's instructions when working on these systems.

DIAGNOSIS OF AIRBAG SUSPENSION SYSTEMS

Diagnosis of airbag systems should be done in accordance with manufacturer's recommendations. However, list below some basic tasks that could identify faults on the system:

- _____
- _____
- _____
- _____
- _____
- _____

- _____
- _____
- _____
- _____

MAGNETO-RHEOLOGICAL DAMPERS

Magneto-rheological (MR) dampers are used by some vehicle manufacturers, they work in a simiar way to an ordinary damper, but they contain MR fluid.

↪ Research magneto-rheological fluid and explain what it is.

_____

_____

_____

_____

Complete the diagram below to show how the oil changes.

_____   _____

_____   _____

Without magnetism   With magnetism

How can the viscosity of the oil be changed?

_____

_____

_____

_____

_____

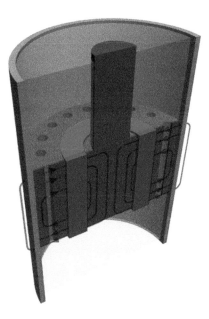

Magneto-rheological fluid damper
operation

MR fluid damper

http://www.carbibles.com/suspension_bible.html

Complete the following paragraph and add these missing words:

current	electric	proportional	viscosity
coil	magnetised	damper	response

The damper has a small _____ coil mounted inside it, the action of the damper forces

magneto-rheological fluid through a magnetised opening in the damper. When a _____ is

sent to the _____ the fluid becomes _____ and its _____ changes

instantly. The oil changes from a fluid state to a semi-solid state that is _____ to the

magnetic field applied to it. With little or no magnetism the fluid flows freely in the _____,

when magnetised the fluid has good dampening characteristics. The system provides an

extremely quick _____ time working in milliseconds.

To operate as efficiently as possible the system will require the use of other sensors around the vehicle, list six of these sensors:

1 _____

2 _____

3 _____

4 _____

5 _____

6 _____

All of these sensors feed information to the electronic control unit. Using algorithms programmed into it, the ECU then adjusts the dampening characteristics to suit vehicle and road conditions.

HYDRO-PNEUMATIC SUSPENSION

Some manufacturers have used hydro-pneumatic suspension on their vehicles for many years. Although the systems can be quite complicated and expensive they can provide a very comfortable ride for passengers and create good handling of the vehicle. They work on a simple principle to do with liquid and gas, what is this basic principle?

_____

The system uses an engine driven pump which may also be used to deliver fluid pressure to which other vehicle systems?

_____

_____

_____

The system usually allows for self-levelling, variable ride height, assisted jacking and little vehicle body roll. Often the height can be controlled by the driver.

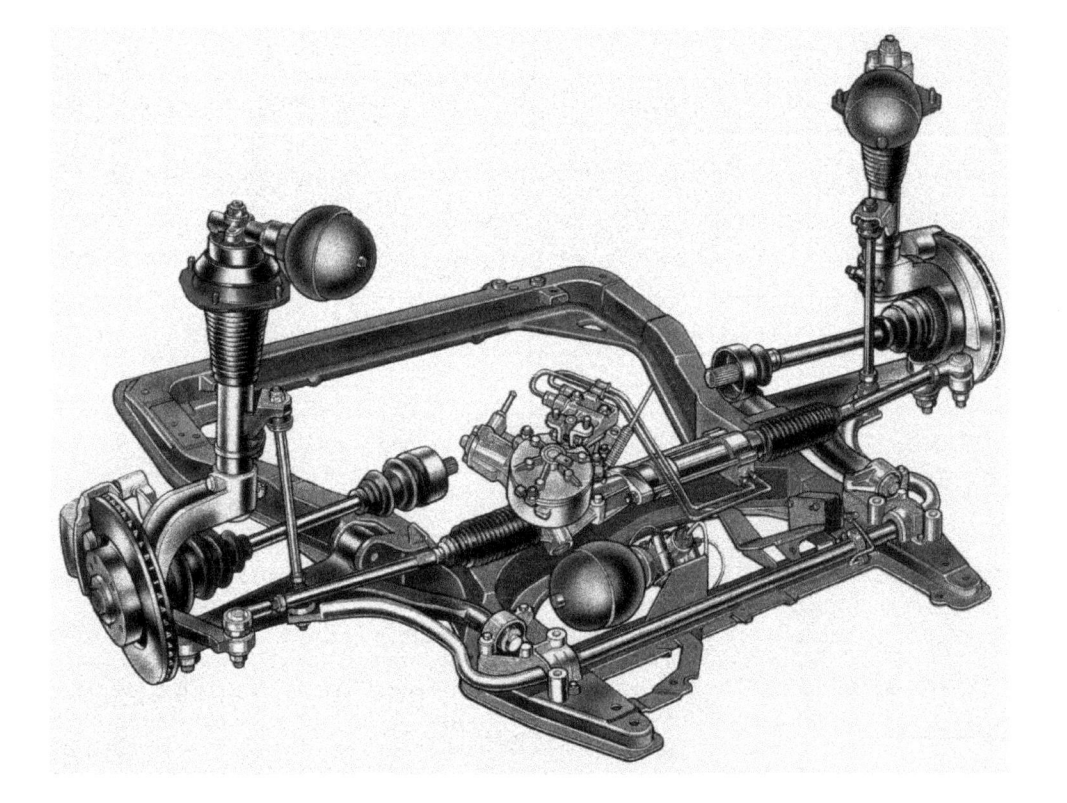

Hydro-pneumatic front suspension system

Hydro-pneumatic suspension pump and reservoir

The pump is driven by the engine and the reservoir must hold enough fluid to allow the spheres to be filled with fluid to fully raise the vehicle.

Some systems use an extra sphere, what is the purpose of this?

_____

_____

Some modern pumps may be electrically driven using sensors from around the vehicle such as steering angle and speed sensors. The pump does not work continuously and only cuts in when pressure in one of the accumulator spheres drops.

Suspension spheres

The suspension spheres, shown to the right, may be found around the vehicle: some are mounted on the strut, some are mounted separately but still connected via pipes, and others are simply to hold an excess of fluid. Wherever they are positioned in the vehicle they work on the same principle.

Suspension spheres

Explain the components of the sphere using the following words:

gas	diaphragm	spring	compressing
green	hydraulic	red	springing
valve	sensor	nitrogen	fluid

The sphere contains a _____. Above the diaphragm shown in _____ is a _____, usually _____ and below the diaphragm shown in _____ is the _____ fluid. The sphere is the equivalent to a _____ in a conventional system, the fluid acts on the diaphragm, _____ the gas which gives a _____ effect.

As the vehicle hits a bump, fluid is forced into the sphere and the diaphragm deflects compressing the gas, when the wheel drops down again then the gas pushes the _____ back out as the diaphragm retakes its shape. The height of the vehicle may be adjusted using a level _____ which on some systems can open a _____ to allow further fluid flow, other manufactures may use an electronic control unit to assist in adjusting the system.

WWW   **http://www.citroenet.org.uk/miscellaneous/hydraulics/hydraulics-1.html**

HYDRO-PNEUMATIC SUSPENSION AND BRAKING SYSTEMS

Manufacturers that use hydro-pneumatic suspension may also use the system to operate the brakes.

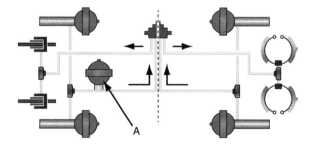

Braking system

The braking system normally employs a separate accumulator sphere (marked 'A' in the diagram on page 155).

What is the purpose of the accumulator marked 'A'?

_____

_____

The rear brakes may take fluid pressure from the rear suspension spheres if the pump fails, why is that a good idea?

_____

_____

_____

Manufacturers that incorporate an electronic control unit may just use the system on the suspension and use conventional steering and braking systems. The system uses a third sphere on each axle. This sphere is switched in and out of circuit by a computer which receives input from sensors measuring steering angle, brake pedal pressure and speed of movement, road speed, body roll and suspension movement.

When the third sphere is 'in circuit' or 'out of the circuit' what affect does this have on the system?

_____

_____

_____

Working on these systems can be dangerous, high pressures and sudden loss of pressure could cause accidents. Follow instructions.

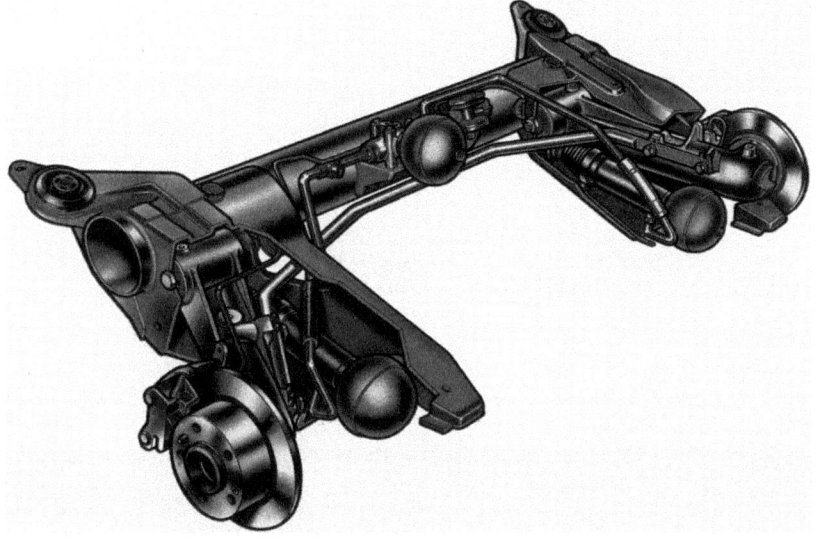

Suspension system incorporating a third sphere

Diagnosis

List some simple ways to diagnose these systems:

● _____

● _____

● _____

● _____

● _____

Use textbooks and the Internet to research a suspension system of your choice and complete a 500 word report on your findings. You should include:

- The key components that make up the suspension system
- A basic overview of how the system works
- The type of vehicles the suspension system is best suited for
- How would you carry out a diagnosis procedure on your chosen suspension system

Multiple choice questions

Choose the correct answer from a), b) or c) and place a tick [✓] after your answer.

1 **It is essential to keep supervisors informed of work progress in order to ensure that the:**

   a) Customer can be informed of potential delays [ ]

   b) Job times will not be exceeded [ ]

   c) Task will be completed on time [ ]

2 **Passive suspension may be considered to be?**

   a) Fully adjustable suspension using sensors [ ]

   b) Non-adjustable suspension [ ]

   c) Only adjustable below speeds of 40mph [ ]

3 **Magneto-rheological fluid thickens as a direct presence of?**

   a) Water [ ]

   b) Oil [ ]

   c) Magnetism [ ]

4 **Hydro-pneumatic suspension uses which gas in the spheres?**

   a) Hydrogen [ ]

   b) Nitrogen [ ]

   c) Carbon neutral [ ]

SECTION 3

Diagnosis and rectification of braking related faults

USE THIS SPACE FOR LEARNER NOTES

_____

_____

_____

Learning objectives

After studying this section you should be able to:

● **Diagnose braking related faults.**

● **Understand methods of finding faults.**

● **Recognise how braking systems work, the parts involved and their operation.**

Key terms

Hygroscopic   Absorbs moisture from the atmosphere.
Braking efficiency   How well the brakes are working.
Regenerative braking   Capturing the vehicle kinetic energy to generate electricity.
Binding brake   A brake that continues to work when the pedal is not pressed and the handbrake is released.
Oversteer   The car turns more than desired by the driver.
Understeer   The car turns less than desired by the driver.

www   http://www.tapley.org.uk

http://sensors-actuators-info.blogspot.co.uk/2009/08/hall-effect-sensor.html

http://www.iihs.org/ratings/esc/esc_explained.html

http://www.carbibles.com/EV_bible.html

http://www.picoauto.com/tutorials/diagnosing-abs.html

DIAGNOSIS AND RECTIFICATION OF BRAKING RELATED FAULTS

Let's remind ourselves of how a master cylinder and a servo unit work.

Look at the diagram with another student, discuss the diagram and see if you can remember why a tandem master cylinder is used. Share your ideas with your classmates.

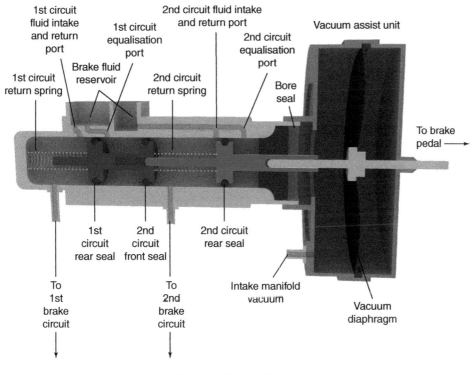

1st circuit fluid intake and return port

1st circuit return spring

Brake fluid reservoir

1st circuit equalisation port

2nd circuit return spring

2nd circuit fluid intake and return port

2nd circuit equalisation port

Vacuum assist unit

Bore seal

To brake pedal →

1st circuit rear seal

2nd circuit front seal

2nd circuit rear seal

To 1st brake circuit

To 2nd brake circuit

Intake manifold vacuum

Vacuum diaphragm

Master cylinder and servo

What is the purpose of the servo?

What is the purpose of the master cylinder?

Modern master cylinders are known as tandem master cylinders; what is meant by this term and why is it important?

How is the vacuum created in the servo?

Describe the correct way to test a servo for operation.

Hydraulic fluid cannot be compressed, but the braking force can be increased or decreased. Use the diagrams opposite to explain how the forces can be increased or decreased.

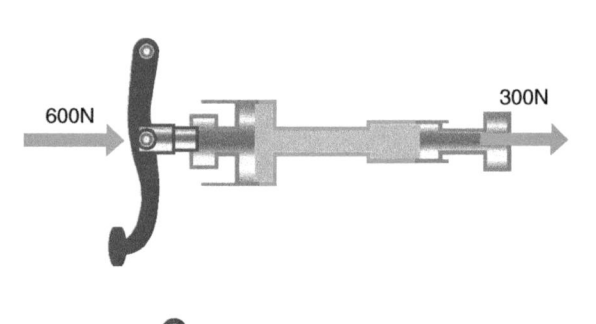

600N → ← 300N

600N → 1200N →

Brake forces

_____

_____

_____

_____

_____

_____

Give an example of where a larger piston is used in the brakes than the piston used in the master cylinder.

_____

_____

_____

 When replacing brake parts make sure the correct parts are fitted. Ensure you compare parts, cross reference part numbers or ask your supervisor.

 Brake components at the wheel can become very hot in operation; take care when working on vehicles that have recently been driven on the road.

BRAKE FLUID

Brake fluid is what makes the system operate correctly. The fluid needs certain properties to work correctly, can you list these properties?

- _____
- _____
- _____
- _____
- _____
- _____

Brake fluid does have a non-desirable property: it is **hygroscopic** which means that the fluid absorbs water from the atmosphere.

 Always use the correct fluid for the vehicle. Brake fluid is graded and given a DOT number – research DOT numbers. Find out what DOT means.

 Many automotive documents use the term 'hydroscopic' but the term is actually 'hygroscopic' – water content in brake fluid is therefore measured using a 'hygrometer'.

Brake fluid may be given a dry boiling point or a wet boiling point, what is meant by these terms?

Dry boiling point – _____

_____

Wet boiling point – _____

_____

Why is the moisture content and boiling point of brake fluid so important?

_____

_____

_____

PADS, DISCS AND CALIPERS

Scenario

A customer has arrived at the garage and wants his brakes checking, no particular fault has been confirmed; the customer has just requested a brakes check. The vehicle could be considered a performance vehicle and therefore has disc brakes on the front and rear.

Disc brake assembly

Make a list below of the checks to be completed in order to carry out an inspection of the brakes:

- _____
- _____
  _____
  _____
- _____
- _____
- _____
- _____
  _____
- _____
- _____
  _____
- _____
- _____
  _____
- _____

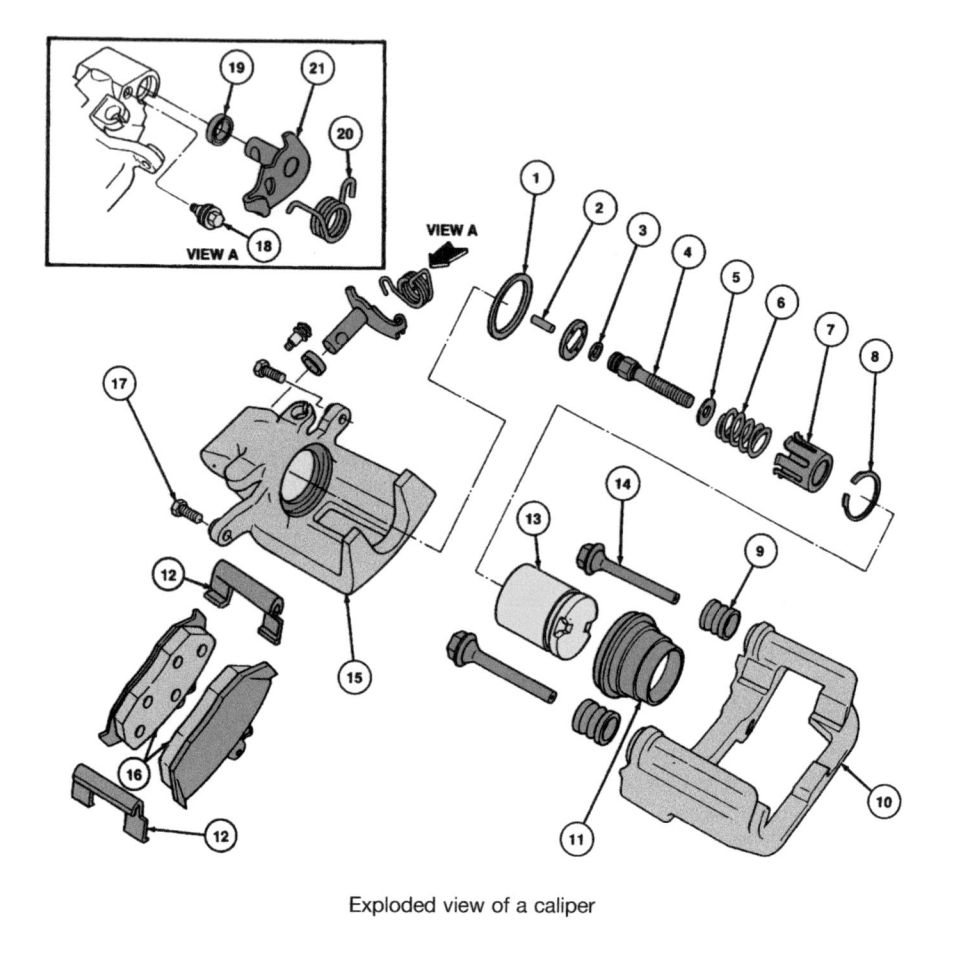

VIEW A

VIEW A

VIEW A

Exploded view of a caliper

The diagram above shows an exploded view of a caliper. Name the numbered components on the diagram.

1 _____   4 _____

2 _____   5 _____

3 _____   6 _____

7 _____   15 _____

8 _____   16 _____

9 _____   17 _____

10 _____   18 _____

11 _____   19 _____

12 _____   20 _____

13 _____   21 _____

14 _____

TIP Caliper slides may need to be lubricated with high melting point grease; use the correct product as recommended by the vehicle manufacturer.

Pads and discs: diagnosis

What would be the most likely cause of the following complaints?

Pulsating brake pedal – _____

_____

Spongy pedal – _____

_____

Grabbing brakes – _____

_____

Binding brakes – _____

_____

Hard pedal – _____

_____

Fluid loss – _____

_____

Noise – _____

_____

_____

Car pulls to one side on braking – _____

_____

_____

TIP Whilst checking for brake faults always make checks to associated components such as steering, suspension and tyres.

Checking disc brake run-out

The image above right shows a Dial Test Indicator (DTI gauge) measuring brake disc run-out. If the disc run-out is being checked what may the customer have complained of?

_____

_____

_____

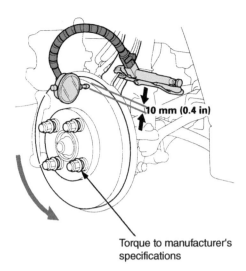

10 mm (0.4 in)

Torque to manufacturer's specifications

Checking brake disc run-out

Brake discs may become warped (or you could say buckled) if they become too hot during braking. It is impossible to see the warp on the disc with the eye so a Dial Test Indicator (DTI) is used to accurately measure the disc. You may see a slightly blue colour on the disc due to heat.

List below the procedure for checking disc run-out with a DTI gauge:

● _____

● _____

● _____

● _____

_____

● _____

● _____

● _____

Measuring brake disc thickness

As brake pads wear, slowly the discs will also wear away. It may be that you will be expected to measure a brake disc to determine its thickness.

The correct tool to use for measuring brake discs is the external micrometer, list the procedure to do this:

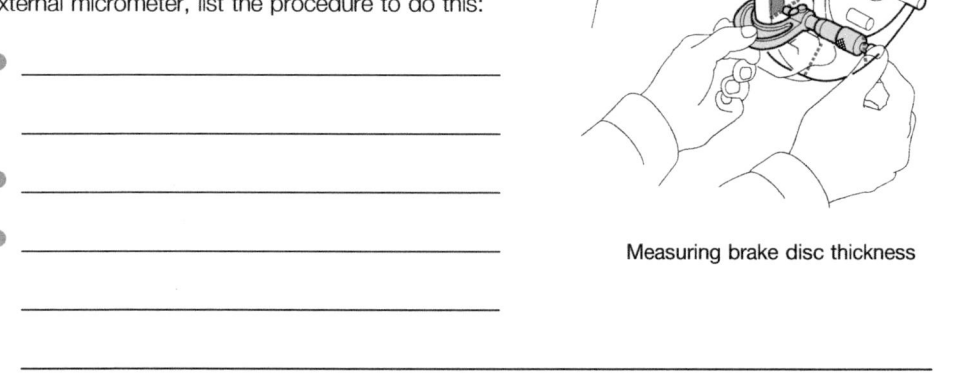

10mm (0.4in)

Measuring brake disc thickness

- _____

  _____

- _____

- _____

  _____

  _____

  _____

- _____

- _____

TIP You should be familiar with micrometers and have a good understanding of how to read one, if you are in doubt complete further research or ask your supervisor.

Enter your place of training or study and measure brake discs for run-out and wear.

Brake drums and shoes: diagnosis

When completing a brake inspection list some components that should be checked on a drum brake assembly:

- _____
- _____
- _____
- _____
- _____
- _____
- _____
- _____
- _____
- _____

Describe the most probable fault based on the customer complaints below and on page 165:

- Rear brakes binding/sticking – _____

  _____

- Rear brakes are squeaking – _____

  _____

- Rear brakes grabbing – _____

  _____

- Brakes judder or vibrate – _____

  _____

● Excessive pedal and handbrake lever travel – _____

_____

Measuring a brake drum for 'ovality' (out of round)

Occasionally a brake drum could over heat, and become oval or out of round in the same way that a disc can overheat and become buckled.

A drum should be inspected for faults but it would be difficult to see if it is oval. There are tools available to measure brake drums for ovality, they may be similar in design to a vernier gauge.

Measuring a brake drum

When drums wear it is common to clean the lip from the drum and continue to use it. However drums will have a discard dimension which is when it becomes too large due to wear. Describe the procedure for measuring the brake drum.

_____

_____

_____

TIP As drums wear excessively they should be replaced. Check manufacturer's data.

ROLLER BRAKE TESTER

Another way to check for brake faults, including brake judder, would be to use a brake roller tester.

How could a brake roller tester be used to determine brake judder?

_____

_____

_____

_____

What does the gauge below indicate?

_____

_____

_____

_____

Brake roller test reading

TIP The brake roller tester can also be used to complete an ovality test for brake drums. This is an excellent way to determine which individual brake is causing the fault.

 Be aware of the brake roller tester in the workshop and cover the rollers when not in operation to prevent people falling in them.

Brake roller test

Brake roller reading

If the brake meter on the brake roller tester was to read even without the brake being pressed, what would this indicate?

_____

The brake tester can now be used to check the efficiency of the brakes, or you could say how well the brakes are working.

Explain how to measure the efficiency of the brakes.

_____

_____

_____

All axles should be tested and the brakes across the axle should read approximately the same; if they don't then this is known as a brake 'imbalance'. An excessive brake imbalance could become dangerous and faults should reported to the customer and investigated.

Why is it important that the front steered wheels read approximately the same?

_____

_____

To check the total braking efficiency now the rear brakes should be checked in the same way as the front. The handbrake (park brake) should be checked for operation and results recorded.

TIP The handbrake should be tested for imbalance. The handbrake should be applied gradually then you should note if the vehicle tends to deviate excessively from a straight line whilst on the rollers, if the vehicle does deviate excessively then the fault should be investigated.

Testing rear brakes

 Research current MOT standards and list the required braking efficiencies.

- _____
- _____
- _____

CALCULATING BRAKING EFFICIENCIES

 You may be expected to calculate braking efficiencies manually, research the formulas to complete these calculations.

To calculate the service brake (foot brake) efficiency:

_____

_____

[ ]

Calculate the example below to show braking efficiency:

A brake test on a vehicle weighing 1800kg the four braking forces add up to 1080kgf:

[ ]

Calculate the parking brake percentage efficiency by:

_____

_____

[ ]

Calculate the handbrake efficiency of a vehicle that weighs 1370kg, the nearside reading is 137kgf

and the offside reading is 115kgf:

[ ]

To calculate the brake imbalance on the steered wheels:

_____

_____

[ ]

Calculate the front imbalance of a vehicle that weighs 1430kg, the nearside braking force is 359kgf and the offside force is 409kgf:

http://www.tapley.org.uk

There are some vehicles that are not suitable to put onto a brake roller tester, which types of cars

are these? _____

_____

_____

The image below shows a decelerometer used to check braking efficiency when the brake roller tester cannot be used.

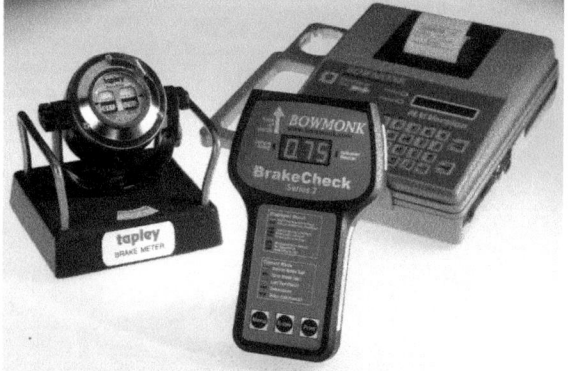

Decelerometer

A customer has asked you to inspect their brakes as they are juddering and don't seem to work as they should; the car is fitted with disc brakes on the front and rear.

Put the following list in order to ensure the task is completed correctly:

- Ask the customer for authorisation
- Check the caliper slides are free
- Ensure the handbrake is not sticking
- Measure the brake fluid water content and record results
- Measure the discs with a dial test indicator if required
- Measure the discs with a micrometer if required
- Pump the brake pedal and have the car road tested
- Put the car on the brake roller tester and record readings
- Raise the car and gather the tools and equipment.
- Refit the wheels and torque up securing bolts
- Remove the front and rear wheels.
- Replace the parts needed to repair the fault
- Speak to the customer to understand the fault
- Use a torque wrench to refit calipers and pad carriers
- Visually check for brake fluid leaks
- Visually inspect the pads and discs for condition and wear

- _____
- _____
- _____
- _____
- _____
- _____

_____
_____
_____
_____
_____
_____
_____
_____
_____
_____

 CHECK Take care when working on electronic handbrakes, some systems will require you to use diagnostic testers during replacement of parts or to recalibrate the system on completion.

Depending on the system there are various sensors that will be needed to ensure the system works correctly.

List some sensors used below:

_____

_____

_____

_____

ELECTRONICALLY CONTROLLED HANDBRAKE SYSTEMS

Electronically controlled handbrake systems are becoming popular with manufacturers. There are different ways that manufacturers operate their systems; some are simply a motor that pulls the handbrake cables tight to activate the handbrake levers on the calipers to lock the brake. Other manufacturers use a motor directly on the brake caliper and an electronic control unit to operate it.

What are the main advantages of having an electronically controlled handbrake?

_____

_____

_____

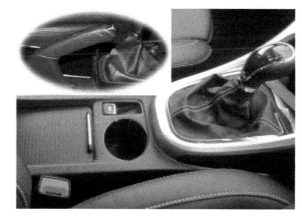

Handbrake lever and switch design

ANTI-LOCK BRAKING SYSTEMS

Antilock braking systems need to determine if the vehicle wheels are skidding, this is calculated by the electronic control unit, using 'slip ratio'.

Research and describe what slip ratio is.

_____

_____

_____

_____

How can percentage slip be measured?

_____

_____

_____

The actual calculation to calculate slip ratio (expressed as a percentage):

$$\frac{\text{Vehicle speed} - \text{Wheel speed}}{\text{Vehicle speed}} \times 100 = \text{slip ratio}\,\%$$

When the slip ratio is 0 per cent the vehicle speed matches the wheel speed exactly (the vehicle is moving), when the slip ratio is 100 per cent the wheels are completely locked. The ideal slip ratio for maximum deceleration is around 10 to 30 per cent.

Can you identify the parts of an ABS system shown below and on page 171?

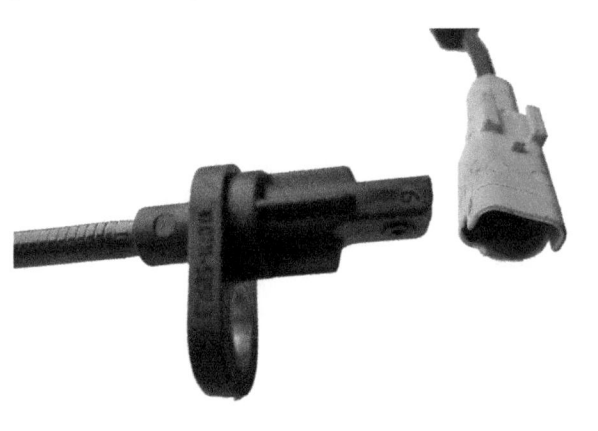

_____

_____

Labels the parts indicated on the diagram below.

Explain how a basic ABS system would work.

_____

_____

_____

_____

_____

_____

_____

How does the ABS system work if the warning lamp stays on during the self-test?

_____

_____

Demonstrate on the graph below how the voltage/frequency would look at low speed and high speed for an inductive sensor. Use red pen for low speed and blue for high speed.

Sine wave pattern

Some manufacturers use Hall sensors as ABS sensors rather than inductive sensors. Draw a Hall effect sensor output on the graph below.

Square wave pattern

http://sensors-actuators-info.blogspot.co.uk/2009/08/hall-effect-sensor.html

Two types of sensor are commonly found – inductive and Hall effect.

How could an inductive (or passive sensor) be recognised?

_____

_____

How could a Hall sensor be recognised?

_____

_____

_____

Hall sensor operation

The sensor generally has three wires, a live feed is sent from the ECU, and they have an earth connection and a signal wire. They contain an amplifier which is why they need a supply voltage. The rotating, alternating, magnetic ring produces a magnetic flux within the sensor element which then amplifies and controls the signal for the ECU to use as wheel speed information. These sensors are very accurate and can recognise even very low speeds.

ABS sensor and hub

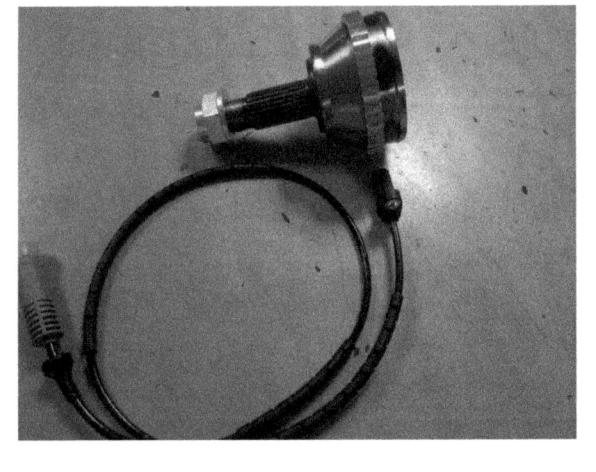

CV joint with reluctor ring and separate sensor

Diagnosis of sensors

List some key steps in diagnosing an ABS sensor fault:

● _____

● _____

● _____

● _____

_____

● _____

● _____

CHECK   Care must be taken when checking ABS sensors – follow manufacturer's instructions to avoid damage to sensitive electrical components.

Modulator operation

When a wheel is showing signs of locking the ECU needs to take action, the ECU cannot complete this task alone and uses part of the hydraulic system, the ABS modulator.

Describe the basic operation of the modulator.

_____

_____

_____

What happens inside the modulator under normal braking?

_____

_____

_____

When the brake begins to lock, what happens inside the modulator?

_____

_____

_____

If the wheel shows further signs of locking what happens in the modulator now?

_____

_____

_____

Depending on how much the wheel speeds up during the pressure release phase, the ECU reverts to the earlier pressure hold or increase phase. If the wheel starts to lock again, the cycle is repeated. The system can operate many times per second and the resulting pressure variations are usually felt at the brake pedal as vibration which is perfectly normal.

ABS modulator

 CHECK Modern technology including the use of Hall sensors and CAN-BUS wiring has allowed ABS technology to move on rapidly allowing many different systems to operate, adding to vehicle safety

Electronic brake force distribution

For many years manufacturers have fitted brake proportioning valves into conventional braking systems, what does a brake proportioning valve do?

_____

_____

_____

_____

_____

_____

What is electronic brake force distribution?

_____

_____

Can this system understand how much load is in the vehicle?

_____

_____

_____

Electronic brake force distribution

Traction control

A vehicle may need traction control to intervene if a wheel is spinning.

If one wheel is spinning faster than another on the same axle how does the ECU know if ABS or TC is required?

_____

_____

If one of the driven wheels starts to spin during acceleration then the ABS/TRC ECU can complete a number of actions, some of these could be?

● _____

● _____

● _____

● _____

TIP Remember, the driver does not press the brake. The system builds up hydraulic pressure in the ABS modulator and applies the brake of the spinning wheel to slow it down.

Brake assist

Something to discuss with your classmates: Imagine driving along the road and suddenly the vehicle in front performs an emergency stop, what is going to be your immediate reaction?

_____

_____

_____

TIP Some manufacturers may refer to this as Emergency Brake Assist.

Brake assist can understand your reactions and intentions.

How can the brake assist system understand that an emergency stop is being performed?

_____

_____

_____

_____

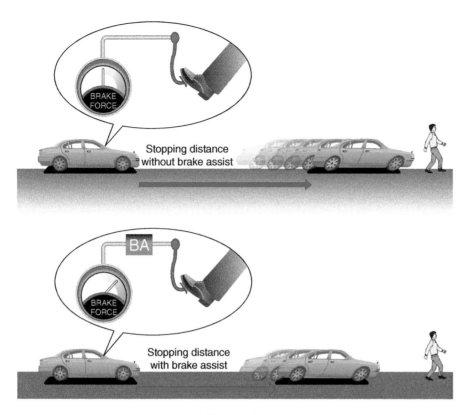

Brake assist

The brake assist programme basically takes over the braking system, allowing maximum braking without wheel lock (ABS is used): a pressure sensor is commonly installed in the master cylinder allowing the system to determine speed and pressure build-up.

VEHICLE STABILITY CONTROL (VSC)

Other names that are used by manufacturers for vehicle stability control could be Electronic Stability Program (ESP), Dynamic Stability Control (DSC).

 www http://www.iihs.org/ratings/esc/esc_explained.html

Research oversteer and understeer, describe the difference.

_____

_____

_____

_____

_____

Vehicle stability control is linked to the ABS system and is an excellent safety system which may be required if the vehicle starts to oversteer or understeer.

Describe oversteer.

_____

_____

Describe understeer.

_____

_____

The vehicle stability control system detects oversteer and understeer situations, and applies a brake on the vehicle to correct its path. Certain sensors are required to complete this task, can you name some?

● _____

● _____

● _____

● _____

Electronic stability control can also be used with further systems such as 'trailer sway prevention' and 'anti-roll over prevention'. Find out what these terms mean.

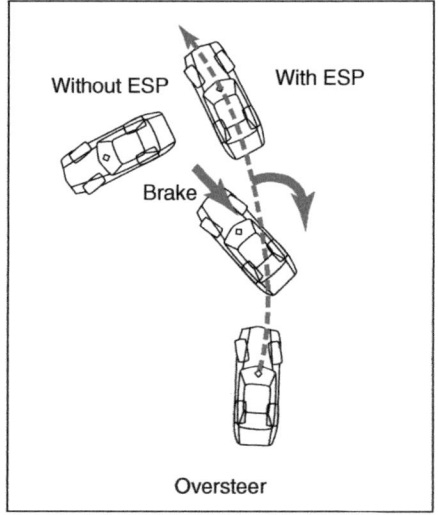 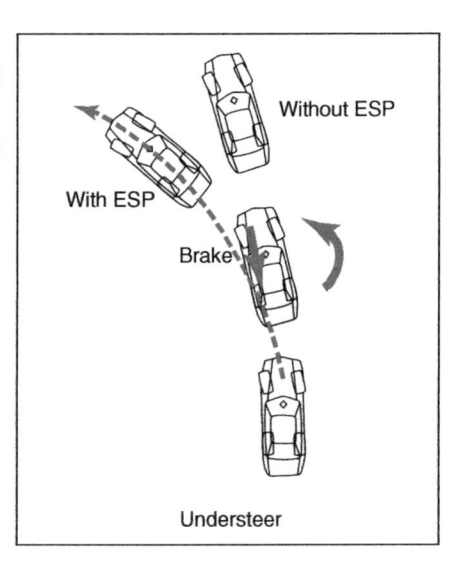

Without ESP   With ESP

Brake

Oversteer

With ESP   Without ESP

Brake

Understeer

Oversteer and understeer

Describe how the system works.

_____

_____

_____

_____

REGENERATIVE BRAKING

Regenerative braking is a term you may have heard of but what does it mean?

_____

_____

_____

_____

Hybrid and electric cars use motors to drive the wheels; these motors can also be used as a generator to charge on board batteries, they are often referred to as a motor/generator (M/G).

When the driver lifts off the throttle pedal and presses the brake, the system knows to recharge the batteries; the vehicle slows down due to magnetic friction caused during the production of the electricity. Hybrid and electric cars still have conventional brakes for heavy braking.

TIP Because hybrids and electric cars slow down through the use of the M/G the conventional brakes have a tendency to stick, if you are servicing these vehicles check caliper slides and pads are free in the carriers.

Extended range hybrid vehicle

http://www.carbibles.com/EV_bible.html

ANTI-LOCK BRAKES AND ASSOCIATED SYSTEMS DIAGNOSIS

Often when these systems operate, the warning lamp may illuminate which is perfectly normal, so the first part of the diagnosis should be to do what?

_____

_____

TIP If you need to road test the car be careful, it may be dangerous to try to duplicate symptoms where vehicle stability control (VSC) is required.

The next stage in the diagnostic process may be to complete a self-test on the system. How is a system self-test performed?

_____

_____

_____

_____

_____

_____

_____

Warning lamps

Scenario

The traction control lamp remains lit, you then check for trouble codes using a diagnostic scan tool, there are eight different trouble codes stored, where would you start your diagnosis?

_____

_____

_____

_____

_____

_____

What do you need to do if the codes don't re-appear?

_____

_____

_____

_____

 TIP Many ABS modulators are not serviceable and parts are not available. They are usually replaced as one unit; often they contain the ECU which may need programming when you buy your new one. Some manufacturers may have a recommended brake bleeding procedure when replacing ABS systems.

Sensors should be checked where possible with a code reader and some, especially wheel speed sensors, can be checked using …?

_____

www **http://www.picoauto.com/tutorials/diagnosing-abs.html**

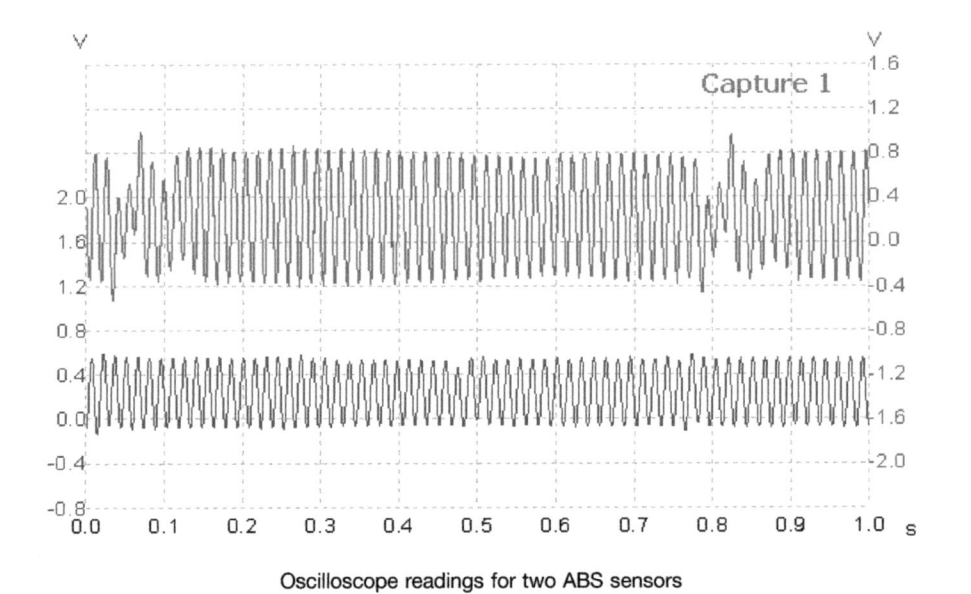

Oscilloscope readings for two ABS sensors

You can see in the diagram above that there is a definite fault shown on one of the ABS circuits. Does this mean that the sensor must be faulty? Explain your answer.

_____

_____

_____

CHECK Always follow manufacturer's instructions when completing diagnostic repair.

Multiple choice questions

Choose the correct answer from a), b) or c) and place a tick [✓] after your answer.

1 **An inductive type wheel speed sensor would give what kind of oscilloscope pattern?**

   a) Square wave [ ]

   b) Sine wave [ ]

   c) Round wave [ ]

2 **Electronic brake force distribution may be used to?**

   a) Reduce braking to the rear brakes in emergency stops [ ]

   b) Distribute force back to the brake pedal [ ]

   c) Lessen the need to press the brake pedal [ ]

3 **A fluctuating needle on a roller brake tester may indicate?**

   a) A buckled disc or drum [ ]

   b) A brake that is not working [ ]

   c) Nothing, this is normal [ ]

4 **A grabbing rear brake on a vehicle fitted with brake drums may be the result of?**

   a) A leaking wheel cylinder [ ]

   b) A snapped handbrake cable [ ]

   c) A split dust cover on the wheel cylinder [ ]

5 **What is the correct repair procedure for an ABS sensor with an infinity resistance reading?**

   a) Replace the reluctor ring [ ]

   b) Adjust the air gap between the sensors and ring [ ]

   c) Replace the sensor [ ]

SECTION 4

Overhauling steering, braking and suspension units

USE THIS SPACE FOR LEARNER NOTES

_____

_____

_____

Learning objectives

After studying this section you should be able to:

● Recognise how systems work, parts involved and their operation.
● Relate customer complaints to faults.

Key terms

Backlash The setting of free play between components.
Overhauling Stripping down and repairing parts and components.
Warped disc A warped disc could be considered to be buckled.

 http://www.carbibles.com/suspension_bible.html

OVERHAULING STEERING SYSTEMS

When checking steering systems, unevenly worn tyres can give clues to faults. You should also pay particular attention to the road wheel and for movement, side to side and up and down. Worn parts can produce free-play which can then be identified.

Side to side movement and up and down movement

What could be faulty if you find movement?

- _____
- _____
- _____
- _____

  _____
- _____
- _____

Steering part

What is the part shown in the figure and what kind of faults may you find on it?

_____

_____

 It may be unusual these days to overhaul a steering rack due to the cost in labour time. You may need to set it up correctly when rebuilding, using a DTI gauge to measure the backlash between the rack and pinion. You may even struggle to find the correct settings; however if you have an old steering rack in your placement or place of study it would be good to strip and rebuild it so that you can understand better how they work.

Name the parts in the steering rack pictures below.

_____

_____

_____

_____

If you are **overhauling** a steering rack inspect the bearings on the pinion, there may be one at each end of the pinion housing.

Check the bearings on the pinon

Check the teeth and seals on the rack and pinion for wear and damage

If the seal inside the rack was to fail where would you expect to see power steering fluid leaking out?

_____

_____

 If you strip a steering rack or steering box take care with oil and greases, wear correct PPE and use manufacturer's specified torque settings and backlash settings where required. Backlash is described as the way that the rack and pinion mesh together, too tight and they could easily wear, too slack and the steering may feel loose.

OVERHAULING SUSPENSION SYSTEMS

Replacing suspension components can be dangerous work if you are unsure about what you are doing ask your supervisor.

Describe the fault on the vehicle opposite.

_____

What would you recommend to the customer?

_____

_____

Once suspension parts have been replaced what should you recommend?

_____

_____

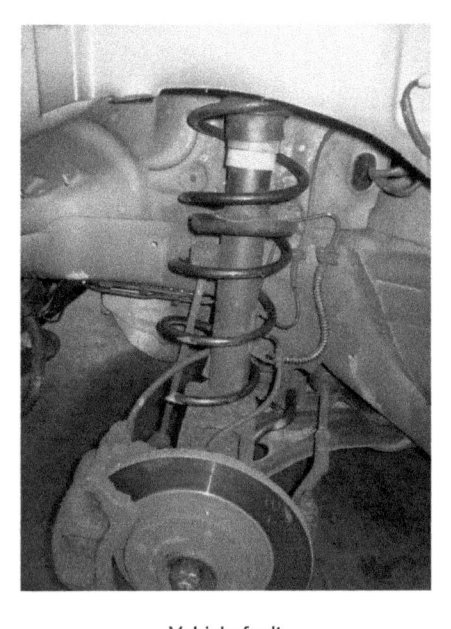

Vehicle fault

Tyre wear can indicate faults with steering and suspension parts.

Explain some possible causes of the tyre wear shown in the pictures below and on page 185.

Tyre faults

_____

_____

Tyre faults

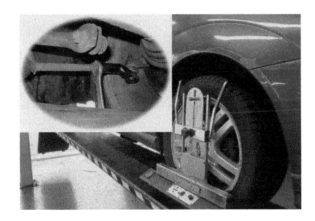

Tracking adjustment

CHECK   If tyres are worn unevenly the vehicle must be repaired and set up correctly at the same time as replacing the tyres or the new ones will wear in exactly the same way.

WWW   http://www.carbibles.com/suspension_bible.html

OVERHAULING BRAKING SYSTEMS

Brake shoe assembly

When inspecting the above brake assembly what should you check on the wheel cylinder?

_____

_____

If the brake shoes were not adjusted enough what might the customer complaint be?

_____

What should be applied between the brake shoe and the back plate?

_____

_____

Why should you inspect the handbrake lever?

_____

Brake disc assembly

When fitting brake pads, what is important about the master cylinder?

_____

_____

Overhauling a brake master cylinder

It is not a common task to repair master cylinders, you may even have difficulty buying repair kits, but if you can find an old one, strip and inspect it so you can get a better understanding of its operation.

Remove the master cylinder from the vehicle, remove the reservoir and if available blank the ports to prevent dirt and moisture entering the master cylinder until you are ready to strip it down. Gather tools as required.

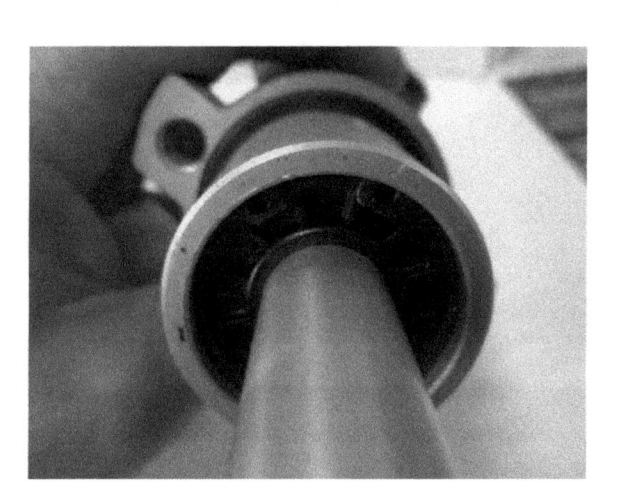

The pistons are generally fastened in with a circlip. The pistons have large springs, be aware when you remove the circlip, parts may jump out with the force of the spring.

Remove the seal to reveal the primary piston, remove this and be careful not to lose parts.

This master cylinder has a pin which drops through from the reservoir port to hold in the secondary piston. This can be removed by easily pushing the piston against the spring pressure.

Investigate the secondary piston, it may simply fall out, or there may be a circlip or a pin locating it. Manufacturers will use different methods.

Inspect the seals for wear, splits and signs of perishing.

Inspect pistons, springs and rods for damage.

Ensure parts are perfectly clean, rebuild in a methodical manner. Ensure all parts are fitted and take care with the strength of the spring. Refit blanking caps where possible to prevent dirt and moisture entering the master cylinder.

CHECK If you are replacing parts to be refitted to a vehicle, the manufacturer may recommend using lubricant on the parts when you refit them. Only ever use the lubricant recommended as you don't want a reaction on the rubber seals.

Workshop task

Enter your place of training or study, replace a brake pipe on a vehicle, ensure the flare is correct and then bleed the system through.

Scenario

A customer wants some parts replacing, – N/S/F coil spring broken, brake discs worn excessively, wiper blades split and N/S/F Flexible brake pipe perished.

Part	Price	Labour
Coil spring	£22.50	1.0
Flexible brake pipe	£10.00	1.7
Brake discs	£40.00	1.3
Fog lamp	£27.00	1.0
Wiper blade	£2.50	-
CV boot	£10.00	1.2
Track rod end	£15.00	0.7
Brake pads	£20.00	1.0
Wheel nut	£3.00	-
Seat belt	£40.00	0.7
Exhaust silencer	£50.00	0.9
Brake fluid	£10.00	-
Wheel alignment	-	1.0

Using the parts list given calculate the cost of repair including labour and VAT, labour is charged at £60.00 per hour and VAT is 20%.

_____ _____ _____
_____ _____ _____
_____ _____ _____
_____ _____ _____
_____ _____ _____
_____ _____ _____
_____ _____ _____
_____ _____ _____
_____ _____ _____
_____ _____ _____

_____

_____

_____

Multiple choice questions

Choose the correct answer from a), b) or c) and place a tick [✓] after your answer.

1　A customer has complained of a juddering through the steering system when braking from high speed, the most likely cause is?

   a) Warped brake discs [ ]

   b) Sticking brake calipers [ ]

   c) The ABS is operating [ ]

2　During an inspection a technician notices power steering fluid leaking from the steering rack gaiters, the most likely cause is?

   a) A missing gaiter clip [ ]

   b) Worn steering rack seals [ ]

   c) Excessive power steering pump pressure [ ]

3　A customer complains that they can feel a vibration through the steering wheel above 50mph, the most likely fault is?

   a) Front wheels out of balance [ ]

   b) Rear wheels out of alignment [ ]

   c) A warped brake disc [ ]

4　A customer complains that the traction control light was illuminated whilst driving on loose gravel, what should you do?

   a) Re-programme the traction control ECU [ ]

   b) Inspect the vehicle but expect the fault to be normal [ ]

   c) Replace the tyres as friction was obviously lost [ ]

5　During an inspection a technician notices a worn track rod end, what else should be quoted with the cost of repairs?

   a) Re-align the steering wheel [ ]

   b) A wheel balance [ ]

   c) A wheel alignment [ ]

PART 5
TRANSMISSION SYSTEMS

USE THIS SPACE FOR LEARNER NOTES

_____

_____

_____

_____

_____

_____

_____

_____

SECTION 1
Clutches 192

1 Clutches 193
2 Clutch faults and diagnosis 194
3 Multiple choice questions 196

SECTION 2
Manual gearboxes 197

1 Introduction 198
2 Manual gearbox: faults 199
3 Multiple choice questions 200

SECTION 3
Four-wheel drive 201

1 Four-wheel drive (4WD or 4 × 4) 202
2 Two-speed transfer box 202
3 Differential locks 203
4 Viscous coupling (VC) 204
5 Multiple choice questions 207

SECTION 4
Automatic gearboxes 208

1   Automatic transmission 209
2   Fluid flywheel 209
3   Torque converter 210
4   Honda six-speed automatic manual transmission (AMT) 213
5   Direct shift gearbox (DSG) 215
6   Epicyclic (planetary) gear trains 216
7   Automatic gearbox mechanical system 218
8   Automatic gearbox hydraulic system 219
9   Oil cooler 221
10 Gear selection (automatic transmission) 221
11 Electronic control 222
12 Continuously variable automatic transmission (CVT) 224
13 Control 226

14 Semi-automatic transmission 227

15 Electronic transmission control 229

16 Testing and fault diagnosis 230

17 Diagnostics: automatic transmission: symptoms, faults and causes 231

18 Multiple choice questions 232

SECTION 5

Drive line 233

1 Drive line shafts and hubs 234

2 Differential 234

3 Limited-slip differential (LSD) 235

4 Inter axle (third) differential 237

5 Drive line, shafts and hubs: symptoms, faults and causes 237

6 Removing and replacing a CV joint boot 239

7 Final drive and differential diagnostics: symptoms, faults and causes 241

8 Multiple choice questions 241

SECTION 6

Electronic components 242

1 Sensors and actuators 243

2 Multiple choice questions 247

SECTION 1

Clutches

USE THIS SPACE FOR LEARNER NOTES

_____

_____

_____

_____

_____

_____

_____

_____

Learning objectives

After studying this section you should be able to:

- Explain the purposes of the clutch.
- Describe the construction and operation of clutch systems.
- Explain the types of faults and their causes in transmission and drive line systems.
- Explain the types of faults and their causes in clutch systems.

Key terms

Pressure plate Bolted to the flywheel and presses the clutch plate to the flywheel face.

Clutch plate/centre plate/drive plate Turns the gearbox input shaft.

Diaphragm spring A dish type spring used in pressure plates.

Clutch release bearing Thrust bearing that pushes on the diaphragm spring fingers.

Clutch slip A fault where torque transmission is reduced.

Clutch judder A fault causing unwanted vibration.

Clutch drag A fault where the clutch does not release fully.

Release mechanism Means of the driver temporarily stopping clutch plate rotation.

Coefficient of friction Ratio of the amount of force needed to cause an object to start to slide.

Slave cylinder Hydraulic cylinder moving the clutch arm.

Torque Turning force measured in Newton metres.

www.automotive-clutches.com

www.luk.com

CLUTCHES

Clutches are important components used in conjunction with conventional transmission systems. They are located between the engine and the transmission. State three purposes of the clutch:

1 _____

2 _____

3 _____

TIP Always tighten clutch bolts in the correct sequence and use a clutch alignment tool and torque wrench.

Clutch disc

The diagram below shows a clutch disc. State three other common names given for this component:

1 _____

2 _____

3 _____

What are the types of materials used for the friction surfaces of a clutch disc?

_____

_____

The springs in the centre of the disc help to provide a smooth take up of drive. They help to reduce crankshaft vibrations and drive train shock.

Riveted between the two friction surfaces there are wave type springs. The purpose of these springs is to cushion the take up of drive and help clutch disengagement, by reducing the chances of the disc sticking to the flywheel.

Clutch pressure plate

The diagram at the bottom of page 193 shows a typical clutch pressure plate. This component has two purposes:

1 _____

2 _____

_____

⚡ Never use compressed air to clean off clutch lining dust. Always use a recommended cleaning fluid and system.

Slave cylinder and release bearing

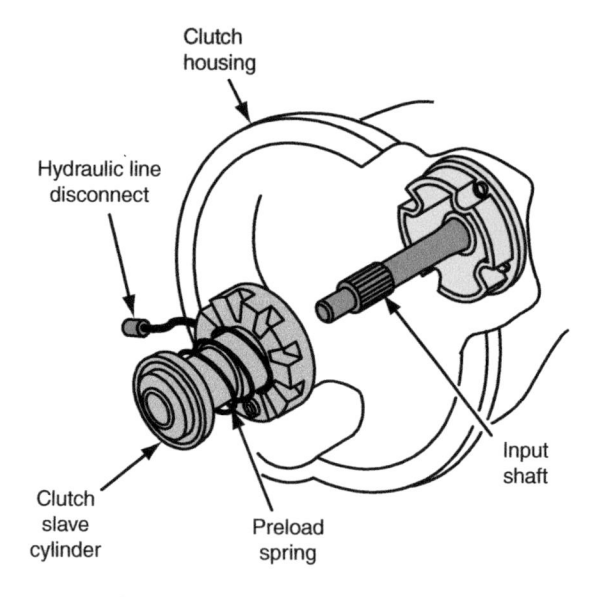

Concentric internal clutch slave cylinder

The release bearing acts upon the fingers of the clutch pressure plate when the clutch is being disengaged. It needs to rotate freely when it is in operation. It is quite common for modern-day vehicles to use a concentric internal clutch slave cylinder as shown in the diagram above.

Dual clutch systems

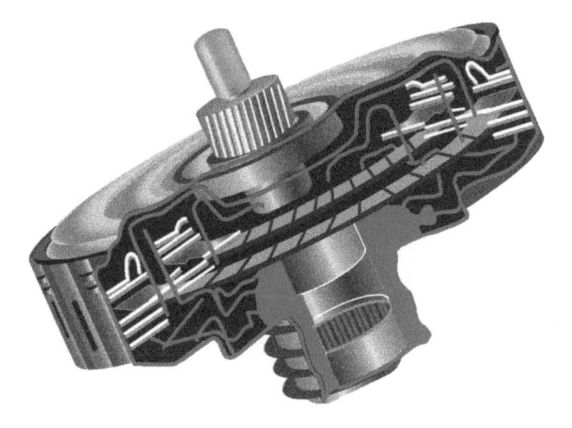

Borg Warner DualTronic™ Clutch System

The type of system shown in the diagram above consists of two clutches in one unit. They are used in transmission systems which have two input shafts and they have automated control. This allows smooth take up of drive, dynamic shifting without any torque interruption. They are used in direct shift gearbox (DSG) transmission systems.

CLUTCH FAULTS AND DIAGNOSIS

A common fault with a hydraulic system is fluid leakage from the slave cylinder. What are the effects of this on clutch operation?

_____

_____

State a clutch fault which would create the need for increasing free movement.

_____

_____

An automatic clutch adjustment is shown in the diagram below. Add the following labels to the diagram:

clutch cable **detent cable groove** **gear quadrant tension**
pawl **clutch pedal** **spring**
gear quadrant

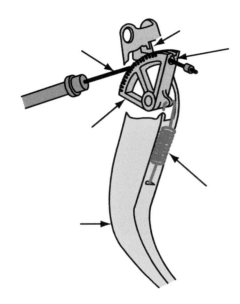

State the effects of a broken quadrant tension spring in the mechanism shown in the image to the left.

_____

_____

_____

State a likely cause for each symptom/system fault listed below. Each cause will suggest any corrective action required.

Symptoms	Faults	Probable causes
Partial or total loss of drive; vehicle speed lower than normal compared with engine speed	Slip	
Difficulty in obtaining gears, particularly first and reverse gears	Drag	
Difficulty in controlling initial take-up drive	Fierceness or snatch	
Shudder and vibration as vehicle moves off from rest	Judder	

Symptoms	Faults	Probable causes
Squeak on first contact with clutch pedal; rattle on tickover	Squeak or rattle	_____ _____ _____ _____
Difficulty in obtaining gears	Spin	_____ _____ _____ _____

Operation	General rules
Avoiding fluid spillage; disposal of waste	_____ _____ _____
Use of clean fluid	_____ _____ _____ _____

Complete the table below in respect of the general rules for efficiency and precautions to be observed during clutch maintenance and repair:

Operation	General rules
Lifting and supporting; preventing distortion	_____ _____ _____
Obtaining correct free play	_____ _____ _____
Correct fitting of **centre plate**	_____ _____ _____
Ensuring component cleanliness	_____ _____ _____

Multiple choice questions

Choose the correct answer from a), b) or c) and place a tick [✓] after your answer.

1   **One possible cause of clutch judder when engaging a gear could be:**

   a) A worn clutch cable [ ]

   b) Badly worn engine mountings [ ]

   c) A broken clutch diaphragm spring [ ]

2   **One possible cause for a clutch not to disengage is:**

   a) Oil on the clutch linings [ ]

   b) The centre plate sticking on gearbox input shaft splines [ ]

   c) A broken centre plate torsional spring [ ]

3   **The springs in the centre of the disc help to:**

   a) Engage the clutch [ ]

   b) Slow down the pressure plate [ ]

   c) Provide a smooth take up of drive [ ]

SECTION 2

Manual gearboxes

USE THIS SPACE FOR LEARNER NOTES

_____

_____

_____

_____

_____

_____

_____

Learning objectives

After studying this section you should be able to:

- Explain how to evaluate and interpret test results found in diagnosing light vehicle manual gearbox faults and compare with manufacturer's specifications and settings.
- Describe the construction and operation of light vehicle manual gearbox.
- Identify light vehicle manual gearbox components.
- Be able to explain the types of faults and their causes in manual gearbox systems.
- Know how to systematically diagnose faults in manual gearbox systems.

Key terms

Interlock mechanism Used to prevent more than one gear being selected at a time.

Spur gear A straight cut gear, which can easily slide in and out of mesh with a similar gear.

Detent Used to ensure positive location of the selected gear.

www.zf.com

http://auto.howstuffworks.com/car-transmission-drivetrain-systems-channel.htm

www.picotech.com

INTRODUCTION

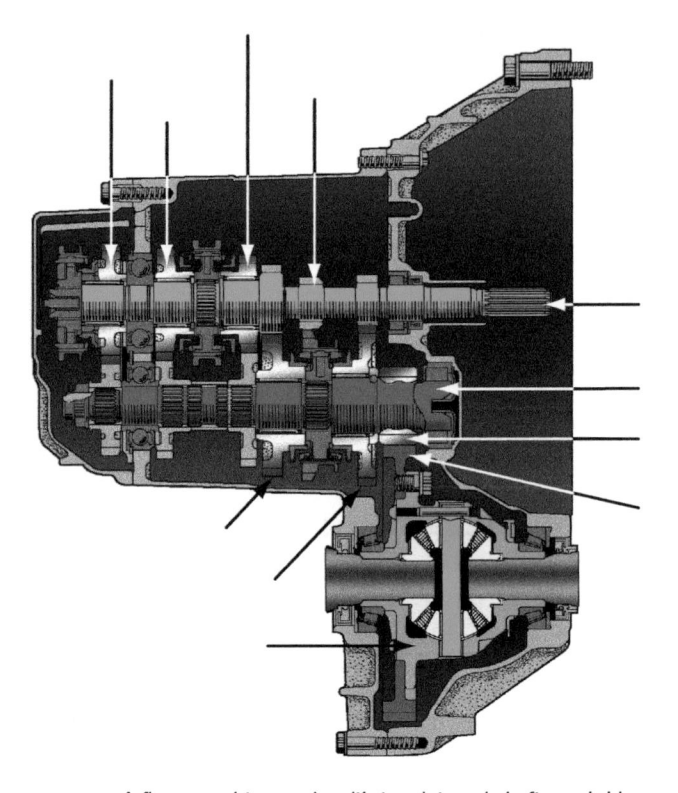

A five speed transaxle with two internal shafts and drive assembly

In pairs correctly complete the arrowed diagram (above) with the following:

drive pinion gear	2nd gear	5th gear	final drive ring gear
output shaft	3rd gear	reverse	input shaft
1st gear	4th gear	differential gear	

The main components of manual gearboxes are gears, which can be either:

_____

With the aid of a partially dismantled gearbox, investigate the types of gears which are commonly used in a manual transmission. List the advantages and disadvantages of using these gears. State the common applications for these gears.

The driver selects the correct gear by moving a hand operated gearstick, which has a neutral position, reverse and either four, five or six gears. The movement of the gearstick is either directly connected to the transmission via rods or a cable mechanism.

Inside a gearbox, when a gear is selected, a selector fork will move a synchroniser hub to engage the selected gear. Power is then transmitted from the engine via the input shaft. A gear at the end of this shaft drives a gear on the countershaft or layshaft. There are a selection of gears on this shaft of varying sizes, which help to create the required gear ratios. These gears in turn drive the gears on the output shaft, also known as the third motion shaft.

A synchroniser mechanism is used to allow a smooth selection of gear inside the transmission.

The system works in a similar way to a friction clutch. It consists of an: _____

_____

How does a synchroniser system work? _____

_____

_____

A detent is used to ensure positive location of the selected gear.

To prevent more than one gear being selected an: _____ is used.

TIP Gears can easily be damaged during the repair of gearbox parts. Take care when fastening components in a vice, use correct equipment where possible.

Carry out diagnostic tests on a gearbox either by operation on road, or dynamometer, or on a partially dismantled gearbox. List the equipment used and faults found.

MANUAL GEARBOX: FAULTS

State a likely fault for each symptom listed below. Each fault will suggest any corrective action required.

Symptoms	Faults
Noisy operation	
Difficulty in obtaining a certain gear	

Symptoms	Faults
Jumping out of gear	
'Sloppy' gear lever action	
Oil leakage from rear of gearbox	
Regular ticking or knocking noise	
Gearbox 'locked up solid', that is shafts will not rotate	
Inoperative speedo/tacho	

Testing and test equipment

Problems associated with the gearbox become evident during testing and checking procedures which include:

-
-
-
-

Multiple choice questions

Choose the correct answer from a), b) or c) and place a tick [✓] after your answer.

1 **A worn synchromesh baulk ring can cause:**

 a) The gears to crash when being engaged [ ]

 b) Excessive vibration on overrun [ ]

 c) Difficulty when disengaging gears [ ]

2 **One possible cause for manual transmission to jump out of gear is a:**

 a) Worn synchromesh assembly [ ]

 b) Weak interlock mechanism [ ]

 c) Worn selector fork [ ]

3 **An incorrectly aligned gear selector mechanism can cause:**

 a) Vibration on acceleration [ ]

 b) Gear changes to be noisy [ ]

 c) Difficulty in engaging a gear [ ]

SECTION 3

Four-wheel drive

USE THIS SPACE FOR LEARNER NOTES

_____

_____

_____

_____

_____

_____

Learning objectives

After studying this section you should be able to:

● Explain the purpose and function of 4-wheel drive systems.
● Explain the function of a third differential.
● Understand the purpose of a torsen unit.
● Describe the construction and operation of a viscous coupling.
● Describe the construction and function of differential locks.

Key terms

Four-wheel drive a system which allows engagement of drive to all four wheels.
Third differential used to prevent windup between the front and rear differential.
Viscous coupling a mechanical device which transfers rotation and torque by means of a viscous fluid.
Limited slip differential a device which limits and controls spin on each of its output shafts.

www.zf.com

www.borgwarner.com/en/transmission/default.aspx

www.picotech.com

**http://www.luk.com/content.luk.de/en/products/clutch_systems
_new/clutch_discs_new/clutch_discs_new.jsp**

FOUR-WHEEL DRIVE (4WD OR 4 × 4)

Four-wheel drive vehicles are either:

1 *Heavy truck type vehicles operating for part of the time off the normal road on uneven, soft or slippery surfaces; or*

2 _____

3 _____

With the simple system shown, four-wheel drive is engaged for use on uneven, soft or slippery surfaces. It should be disengaged for normal road use. Why is disengagement necessary?

_____

_____

_____

_____

_____

Layout

Complete this drawing to show the front and rear final drive gears and show how the drive to the front and rear axle can be disconnected.

Front axle

Front pinion

Clutch housing

Gearbox

Gearbox mainshaft

Idler gear

Sliding joint

Simple transfer box

Output gear

Universal joint

Rear pinion

Rear axle

TWO-SPEED TRANSFER BOX

Complete the drawing and labelling of the two-speed transfer box below and describe its operation in low and high ratio.

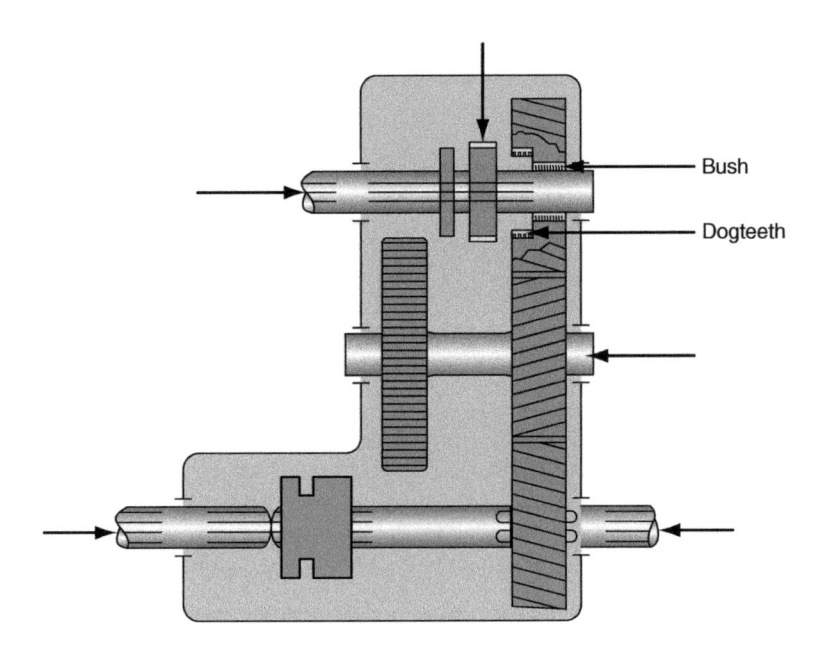

Bush

Dogteeth

Operation (two-speed transfer gearbox)

_____

_____

_____

_____

_____

How does a two-speed transfer box affect the transmission gearing as a whole?

_____

_____

_____

_____

Complete the drawing below to show how the front and rear propeller shafts can be driven through a differential:

State the reason for using a third differential:

_____

_____

_____

 To obtain full benefit of four-wheel drive when operating off the road it is usual practice to use a differential lock on the third differential.

DIFFERENTIAL LOCKS

If one driving wheel of a vehicle encounters a soft or slippery surface, the wheel will spin and the torque required to drive it will be negligible. Because of the action of the differential the other non-slipping wheel will receive the same negligible torque. The vehicle will therefore be immobilised.

The differential lock, as the name implies, locks the differential and allows maximum tractive effort allowed by the road surface to be utilised at each wheel.

Suggest three possible applications of the differential lock:

1 _____

2 _____

3 _____

What safety precautions should be taken when working on the transmission of a vehicle fitted with a differential lock?

_____

_____

_____

_____

_____

_____

_____

If two elements of a differential are locked together, the entire assembly is locked. The device shown on the next page operates on this principle.

Complete the diagram of the differential lock disengaged. Complete diagram in (b) below to show the differential lock engaged, and explain how the action locks the differential:

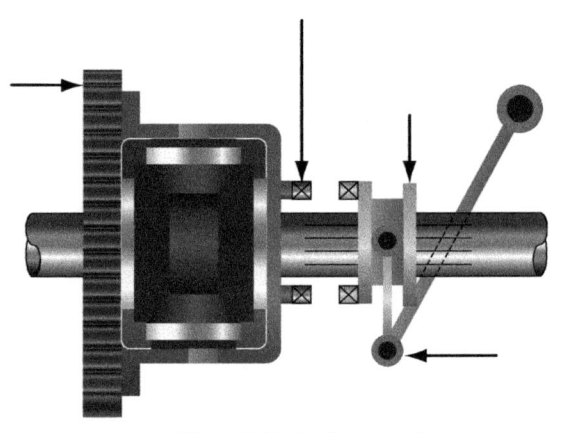

Differential lock disengaged

(a)

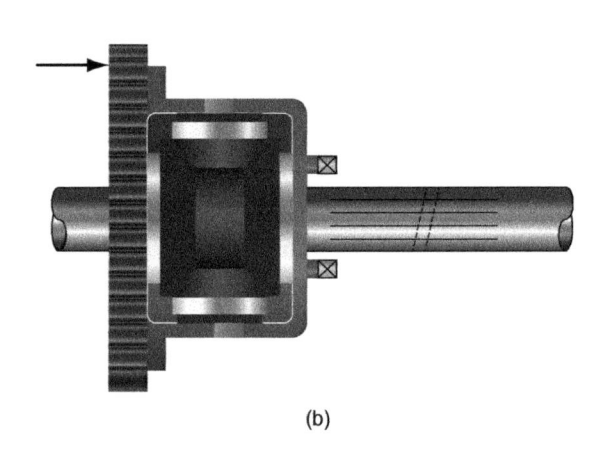

(b)

TIP Accurate measuring tools such as DTI gauges may be required to set up differentials correctly – ensure they are in good condition and read accurately.

Operation

Describe how the differential locks operate.

_____

_____

_____

_____

VISCOUS COUPLING (VC)

Four-wheel drive operation under all conditions can be achieved by the use of limited slip devices. The system shown in the next diagram employs viscous couplings in the centre (third) differential and in the rear differential. These couplings control wheel spin and greatly improve traction and road holding in all drive conditions without the need for the engagement of manual differential locks by the driver.

On a car application it is usual to divide the driving torque unequally between front and rear wheels.

Give a typical percentage torque split and state the reasons for this.

Torque split: Front _____   Rear _____

_____

_____

Where in the transmission is the torque split achieved?

_____

How is the drive transmitted to the front propeller shaft? _____

_____

_____

Complete the labelling on the diagram shown below:

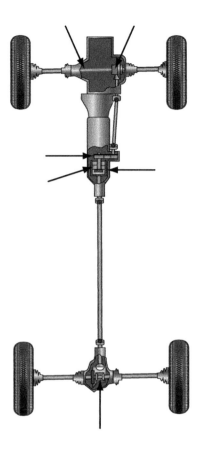

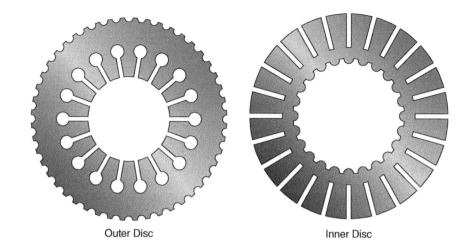

Outer Disc   Inner Disc

The inner discs are splined on to an inner carrier shaft and the outer discs are splined on to the inside of an outer housing. A small clearance or gap between the discs is maintained by interposed spacer rings. The gap between the discs is filled with a high viscosity silicone fluid.

Torque transmission through a VC is based on the transmission of shearing forces in the fluid.

Viscous couplings, as already stated, will control differential spin. In what other capacity are they employed in vehicle transmissions?

_____

_____

Viscous coupling (VC): operation

The structure of a viscous coupling is similar to that of a multi-plate clutch. The coupling consists of a number of inner and outer discs.

Complete the drawing to show the inner and outer discs in the simplified layout.

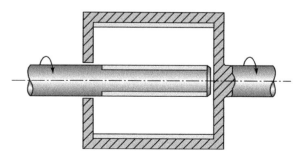

How is torque transmitted from the inner shaft to the outer housing?

_____

_____

_____

_____

Viscous coupling and differential

The viscous coupling, or viscous control, when used in conjunction with a differential will limit or control spin on shafts being driven via the differential, for example, front/rear propeller shafts or drive shafts.

The simplified drawing below shows a VC incorporated into a final drive/differential unit.

Label the drawing and explain how the VC limits drive shaft spin.

_____

_____

_____

_____

_____

_____

_____

_____

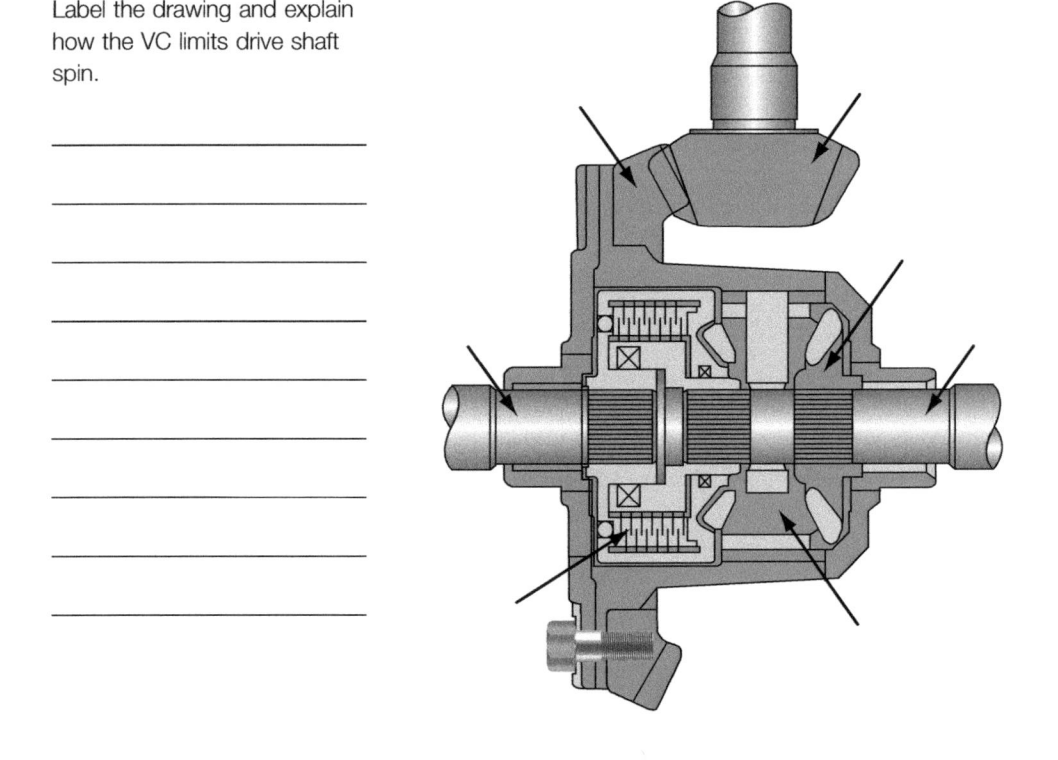

Torsen unit

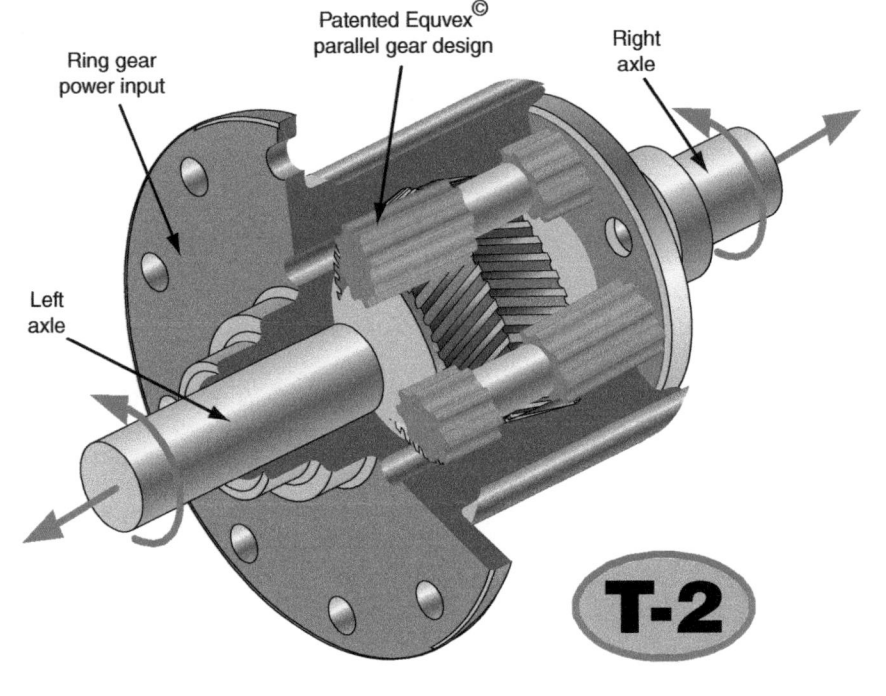

Ring gear power input

Patented Equvex© parallel gear design

Right axle

Left axle

T-2

A torsen torque-sensitive LSD

Some vehicle applications use gear based units, which are often called torque bias or torque sensing (torsen) units. These types of units multiply the torque available from the wheel that is losing traction and sends the torque to the wheel with the better traction. This action is carried by the resistance between the gears that are in mesh.

State the advantages of using this type of limited slip differential (LSD) over the clutch type:

_____

_____

_____

Centre differential

The centre differential shown below is an **epicyclic gear** set. Complete the drawing by adding a VC which would control spin between front and rear propeller shafts.

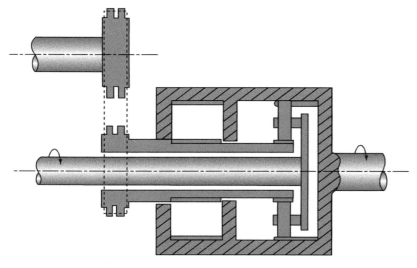

Centre differential

How does this differential provide the 2:1 torque split?

_____

_____

_____

Choose the correct answer from a), b) or c) and place a tick [✔] after your answer.

1 **Which one of the following can be used a third differential on a 4WD system?**

   a) Viscous coupling [ ]

   b) Synchroniser hub [ ]

   c) Transfer box [ ]

2 **Which component, when engaged, allows maximum tractive effort to be utilised by each wheel?**

   a) Transfer box [ ]

   b) Differential lock [ ]

   c) Differential [ ]

3 **Which one of the following best describes a torsen unit. It is a:**

   a) Differential lock [ ]

   b) Type of gear [ ]

   c) Limited slip differential [ ]

SECTION 4

Automatic gearboxes

USE THIS SPACE FOR LEARNER NOTES

_____

_____

_____

_____

_____

_____

Learning objectives

After studying this section you should be able to:

- Explain how to evaluate and interpret test results found in diagnosing light vehicle automatic gearbox and compare with manufacturer's specifications and settings.
- Describe the construction and operation of light vehicle automatic gearboxes.
- Identify light vehicle automatic gearbox components.
- Be able to explain the types of faults and their causes in automatic gearbox systems.
- Know how to systematically diagnose faults in automatic gearbox systems.

Key terms

Torque convertor   A type of fluid coupling which multiplies torque.
Epicyclic gear train   A set of intermeshing gears consisting of the sun, planet and annulus.
Kinetic energy   The amount of energy an object possess due to its motion.
Annulus   A ring gear with internal teeth which mesh with other gears.
Stall test   Is the method used for testing a torque converter coupling.
CVT   Continuously variable transmission, which has an infinite number of gear ratios.

WWW   **www.zf.com**

www.borgwarner.com

AUTOMATIC TRANSMISSION

An automatic transmission system fulfils exactly the same requirements as a manual transmission in that it:

1 Multiplies engine torque to suit varying load and speed requirements.

2 _____

3 _____

In addition to these functions the automatic transmission provides automatic gear changing, that is, the gearbox ratios are selected automatically to meet the speed and load requirement of the vehicle.

What is the difference between fully automatic transmission and semi-automatic transmission?

_____

_____

_____

TIP Automatics are complicated in design and operation, if in doubt ask your supervisor.

Most fully automatic transmissions, however, have a manual override facility with which the driver can dictate when gear selections are made. Another feature of modern automatic transmission is a 'mode' selector — for example, sport or urban; economy or power drive programmes. These settings affect gear selection according to the way in which the vehicle is being operated.

Explain the difference between fixed ratio and stepless transmission.

_____

_____

FLUID FLYWHEEL

A fluid flywheel is a hydraulic coupling which is used as an automatic clutch in the transmission system of a vehicle. The unit consists of two main elements: an impeller or driving member, and a rotor or driven member. The unit is almost completely filled with fluid. As the engine, and hence the impeller, rotate, the fluid begins to circulate; this transfers torque to the rotor and consequently the gearbox input shaft.

Name the type of gearbox used with this type of coupling:

_____

Add arrows to the drawing on the next page to show the direction of fluid circulation.

State the reason why this type of coupling is generally not used with a normal gearbox.

_____

_____

_____

Name a modern vehicle using a fluid flywheel:

_____

_____

In most modern fluid flywheel applications the fluid flywheel is used in conjunction with a lock-up clutch. State the reason for this.

_____

_____

_____

A fluid flywheel engages automatically, thus simplifying driving technique, and it serves as an effective transmission damper.

Fluid flywheel assembly

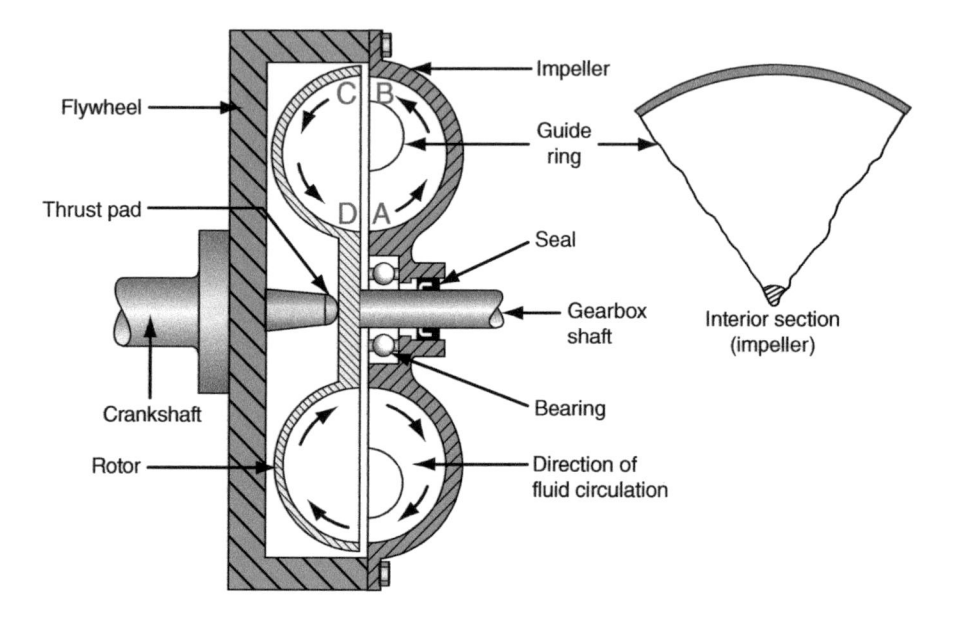

Flywheel

Thrust pad

Crankshaft

Rotor

Impeller

Guide ring

Seal

Gearbox shaft

Bearing

Direction of fluid circulation

Interior section (impeller)

Examine the interior of a fluid flywheel assembly and complete the section of the impeller, shown on the right above, by showing the arrangement of the vanes.

Why does the fluid flow from:

1 A to B? _____

_____

2 C to D? _____

_____

In what other direction is the fluid moving?

_____

Describe how the fluid forces the rotor round.

_____

TORQUE CONVERTER

A cutaway of a modern torque converter

The function and action of a torque converter are somewhat similar to those of a fluid flywheel, but with (in its simplest form) the addition of another fixed, bladed member. The advantage of this arrangement is that when 'slip' is taking place a torque multiplication is obtained.

State the type of gearbox normally used in conjunction with a torque converter.

_____

A simple line diagram representing a two element torque converter is shown to the right, indicating the path of oil flow.

One significant constructional feature of a torque converter is the shape of the vanes in both the impeller and turbine. The greater the change in fluid direction after striking the turbine vanes, the greater will be the force on the turbine. This change in fluid direction is achieved by curving the vanes.

State the reason why it is necessary for the fluid to pass through a reaction member before going back into the impeller:

(a)

(b)

(c)

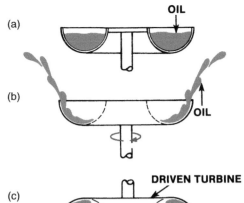

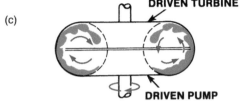

OIL

OIL

DRIVEN TURBINE

DRIVEN PUMP

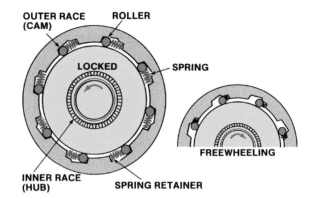

OUTER RACE (CAM)   ROLLER

LOCKED   SPRING

FREEWHEELING

INNER RACE (HUB)   SPRING RETAINER

State the reason why the reaction member (stator) is mounted on a freewheel (overrunning clutch) which is shown above.

_____

_____

_____

_____

Examine a torque converter and complete a sketch showing the shape of the vanes.

The converter shown in the photo on page 212 is a 'three-element' or _____ unit, the

maximum torque multiplication for such a unit is:

_____.

State when maximum torque multiplication occurs.

_____

_____

_____

Name the component parts on the torque converter assembly shown below.

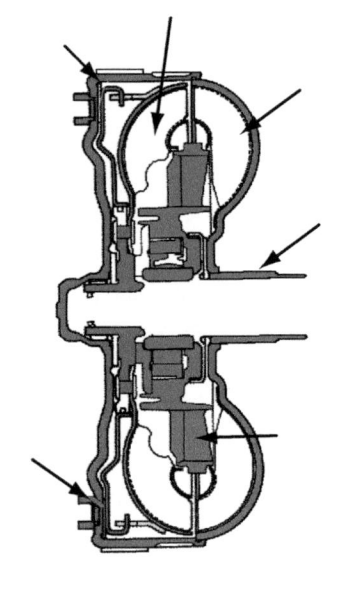

Coupling characteristics

The graph in (a) shows the efficiency output curve for a constant input speed. For the fluid flywheel, complete the graph in (b) to show the curve for a torque converter:

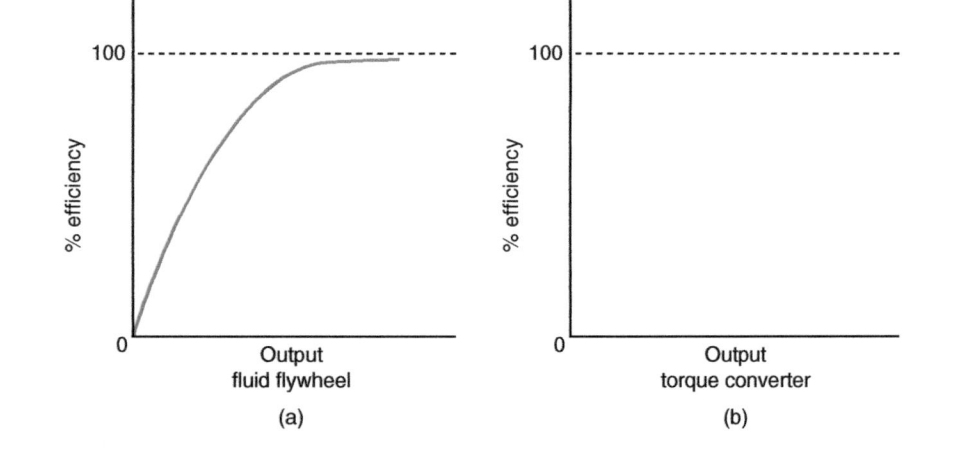

Considering the formula kinetic energy = $\frac{1}{2} mv^2$, state the relationship between fluid speed and kinetic energy.

The velocity and hence the kinetic energy of the fluid are increased as it moves to the outer radius of the impeller. State how the kinetic energy is converted to a force which produces a driving torque in the turbine or output shaft.

Some energy is lost because of conversion to heat energy, that is, the fluid is heated as it is made to work on the turbine. Under which operating condition is maximum heat generated?

Lock up torque converter

To help reduce slip between the impellor and the turbine in a torque converter, by up to 10 per cent, a clutch assembly is used. This type of system is known as a lockup torque converter.

A clutch is incorporated between the impellor and the turbine.

What methods of engagement are used for the lock up clutch?

The lock up of the clutch is controlled by means of hydraulic valves, which are in turn controlled by the engine ECU, dependent upon the operating conditions.

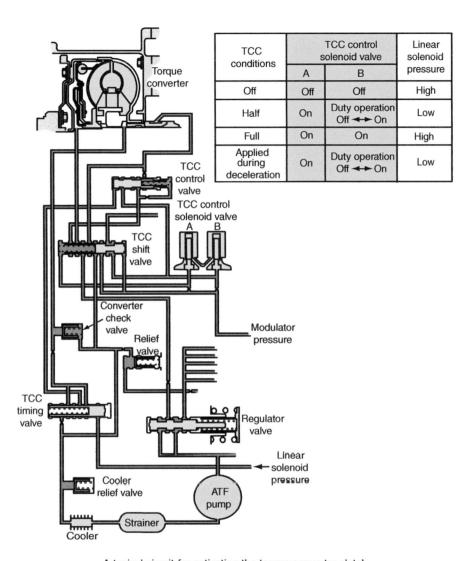

TCC conditions	TCC control solenoid valve		Linear solenoid pressure
	A	B	
Off	Off	Off	High
Half	On	Duty operation Off ←→ On	Low
Full	On	On	High
Applied during deceleration	On	Duty operation Off ←→ On	Low

A typical circuit for activating the torque converter clutch

System case study

The Honda system is a traditional six-speed manual gearbox which has automatically controlled gear shifting. A clutch is still used, with the clutch pedal being replaced by an electromotor that engages and disengages the clutch without driver involvement.

Two electromotors are incorporated to shift the gears; one for selecting the gears and the other for shifting the transmission into gear. The two motors are housed in a unit called the gear change actuator.

A reprogrammable transmission control unit (TCU) controls all three electromotors and is integrated with the clutch actuator.

The AMT consists of the following components:

- _____
- _____
  _____
  _____
- _____
- _____
- _____
- _____
- _____

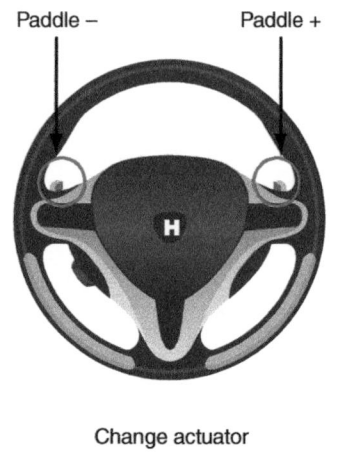

Paddle −   Paddle +

Change actuator

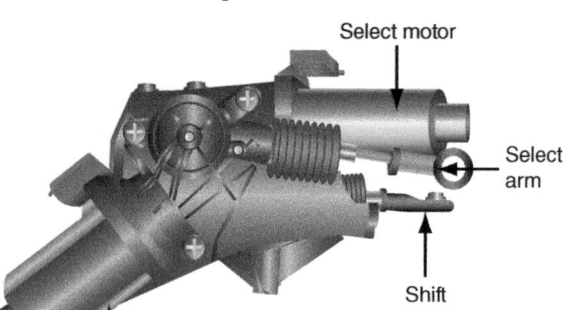

Select motor

Select arm

Shift arm

Shift motor

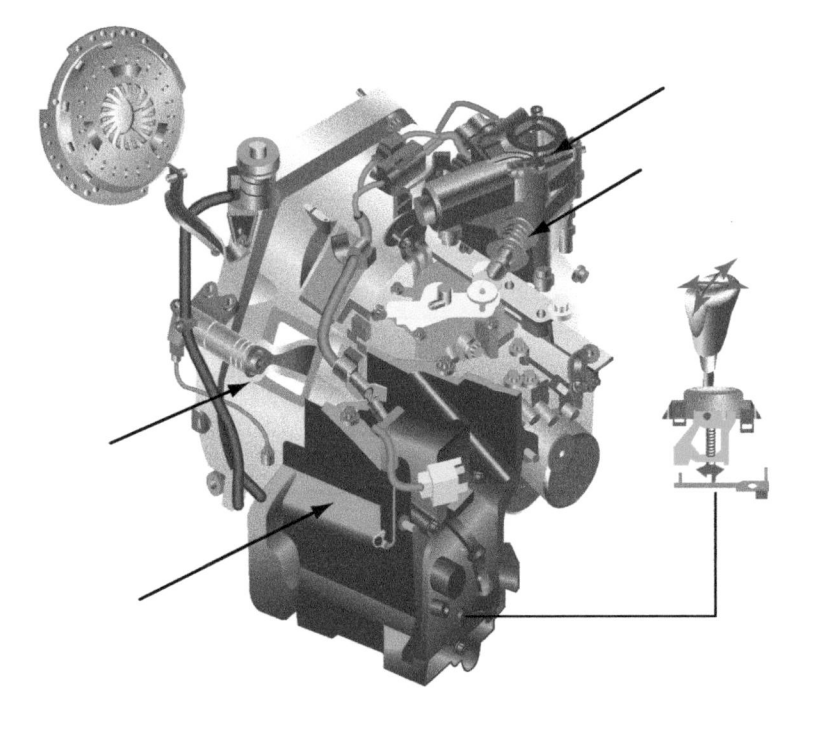

Correctly label the diagram with the following:

clutch actuator with built-in TCU   **self adjusting clutch**   **gear change actuator**

shift lever   **standard Honda six-speed transmission**

The system consists of two settings:

Manual mode: _____

_____

_____

Automatic mode: _____

_____

_____

DIRECT SHIFT GEARBOX (DSG)

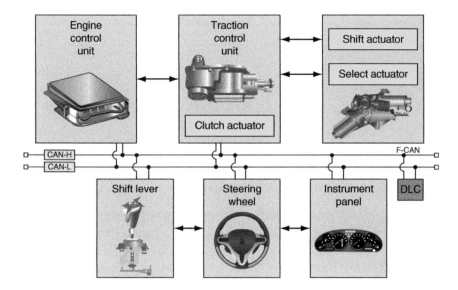

The diagram above shows the major components for the system and how they link to and utilise the controller area network (CAN) for transmitting and sharing information.

The manufacturer's dedicated computerised diagnostic equipment is required to carry out the following service procedures:

- **Replacing the transmission**
- **Replacing the clutch actuator**
- **Replacing the gear change actuator**
- **Bleeding and filling of the hydraulic system**

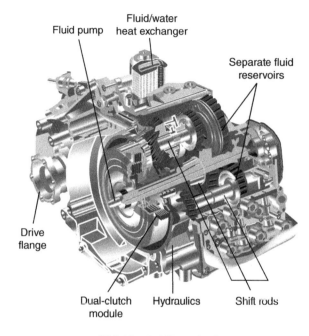

DSG (direct shift gearbox)

This dual-clutch gearbox basically consists of two independent gearbox units. With dual-clutch technology –two wet clutches in a common housing – both gearboxes are connected. Research DSG gearboxes to find what the two clutches operate.

_____

_____

215

Each gearbox unit is assigned an output shaft that applies the torque to the driven wheels via the differential gear.

The principle of gear changes in the DSG is as follows: when one gear is engaged, another gear is always pre-selected. If a gear change is approaching, one clutch opens within hundredths of a second while the other closes. The electronic control unit operates solenoids and ensures that gear changes take place faster and more accurately than would be possible manually. In this way, the change of gear is very smooth without interruption to the driving. On some models, paddles can be attached behind the steering wheel or a lever housed in the centre console. The system will also allow for fully automatic operation.

The decision as to which gear should be engaged is taken by the control unit based on the position/operation of the accelerator, engine speed and vehicle speed. If the accelerator pedal is pressed and the vehicle is accelerating, the next-highest gear will be pre-selected before the gear change point is reached. If the accelerator is not operated, perhaps if the vehicle is in overrun, the next-lowest gear will be pre-selected.

EPICYCLIC (PLANETARY) GEAR TRAINS

Epicyclic gear trains provide, in a very compact manner, various ratios and directions of rotation. This is achieved by holding certain members and applying power to one of the other members. The members are held for gear engagement purposes by brake bands or multi-plate clutches that are normally actuated hydraulically.

List some common motor-vehicle applications of epicyclic gearing:

1 _____

2 _____

The compactness of the epicyclic gear train is an advantage. State one other advantage of epicyclic gearing.

_____

_____

_____

Complete the labelling for the simple epicyclic gear train shown below.

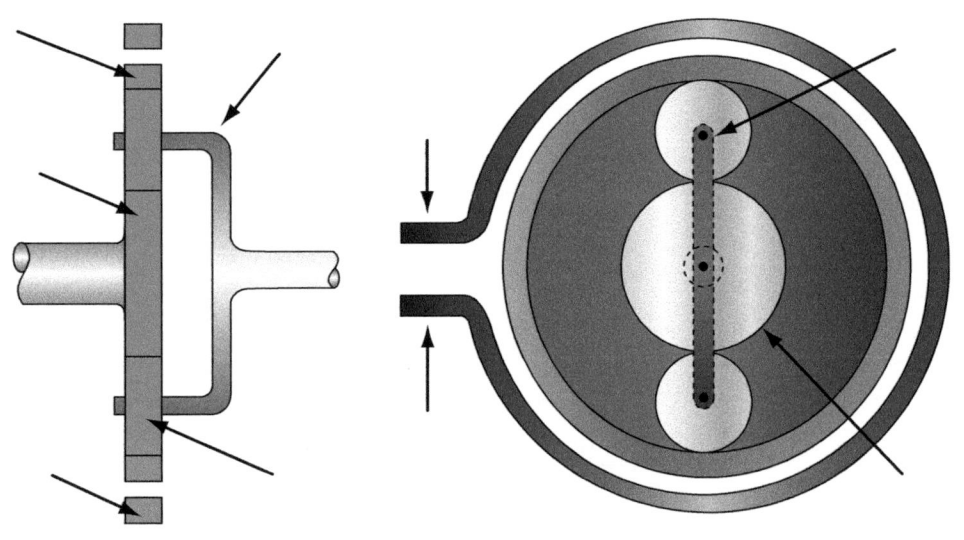

Investigation

Examine a simple epicyclic gear train and study the three combinations shown below.

Describe beneath each drawing the relationship between input and output in each case.
Add arrows to the drawings to indicate the direction of rotation of the wheels.

Brake on

Input sun wheel–output planet carrier

_____

_____

_____

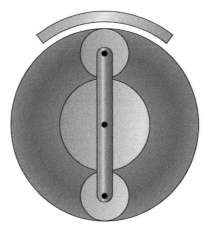

Brake off

Input sun wheel–output annulus–planet
carrier locked

_____
_____
_____
_____

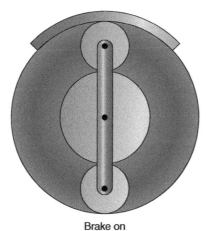

Brake on

Input planet carrier–output sun wheel

_____
_____
_____
_____

Describe how it is possible to increase the number of gear ratios available using epicyclic gearing.

Compound epicyclic gear trains

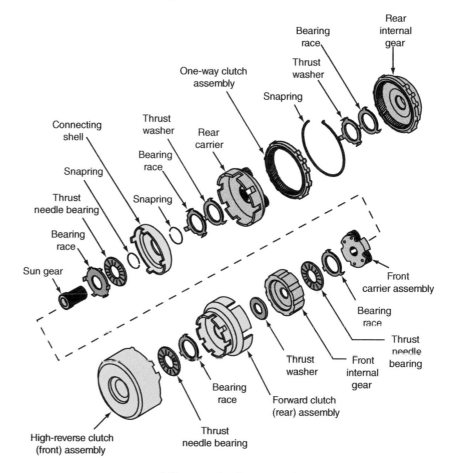

A Simpson planetary gear set

This is a more complex and sophisticated type of epicyclic arrangement than the simple epicyclic gearing mentioned earlier. It has the advantages of being able to obtain a greater number of forward ratios, and reverse. The Simpson gear train is the most commonly used compound planetary gear train.

A 'two-element' epicyclic gear train, as used in some automatic gearboxes, is shown below; this gear set will provide three forward gears and one reverse gear. One important feature to appreciate is the fact that the planet carrier is mounted on a freewheel or one-way clutch; this will only permit the carrier to revolve clockwise (as seen from the drawing). Examine such a gear set and in the spaces provided describe its operation in each gear.

_____

_____

1st gear, input to small sun gear

_____

_____

2nd gear, input to small sun gear – large sun gear locked

_____

_____

2nd gear input to small sun gear – large sun gear locked

Top gear, small sun gear and large sun gear locked together – input through both

_____

_____

Reverse gear, input to large sun gear – small sun gear free – freewheel locked

_____

_____

AUTOMATIC GEARBOX MECHANICAL SYSTEM

The drawing on the next page represents the complete mechanical system in an automatic gearbox. The various ratios are obtained through a compound epicylic gear set such as outlined earlier.

Although the automatic gearbox arrangement shown is an early model, it is a relatively simple example and thus ideal for the purpose of gaining an understanding of automatic gearbox operation.

Study the control system for the gear set and indicate, on the table below the diagram on page 219 and on the diagram, which clutches and brake bands are operative in each gear.

State:
1 the conditions under which the freewheel is operating, and
2 the effect of this action.

_____

_____

State the effect of the application of the rear brake band in first and second gears.

_____

State the reason why the rear brake band is applied in reverse gear:

_____

_____

_____

Mechanical system fully automatic gearbox

Complete the labelling on the drawing.

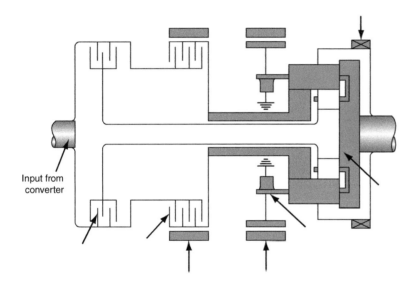

Input from converter

	1st gear	2nd gear	Top gear	Reverse	Neutral
Front clutch					
Rear clutch					
Front brake band					
Rear brake band					

The hydraulic system in the automatic gearbox serves three basic purposes. It:

1 **Maintains a pressurised supply of fluid to the torque converter.**

2 **Provides gearbox lubrication.**

3 **Actuates hydraulic servos during multi-plate clutch and brake band operation.**

A typical valve body

Name, state the purpose and describe the operation of the component shown here.

_____

_____

_____

_____

_____

_____

_____

A gear type oil pump

Name the types of fluid seal used and their particular application in an automatic transmission.

_____

State the reason why the fluid level changes when the selector lever is moved.

_____

_____

Hydraulic servo arrangements for brake band and multi-plate clutch operation are shown below. Complete the labelling on the drawings.

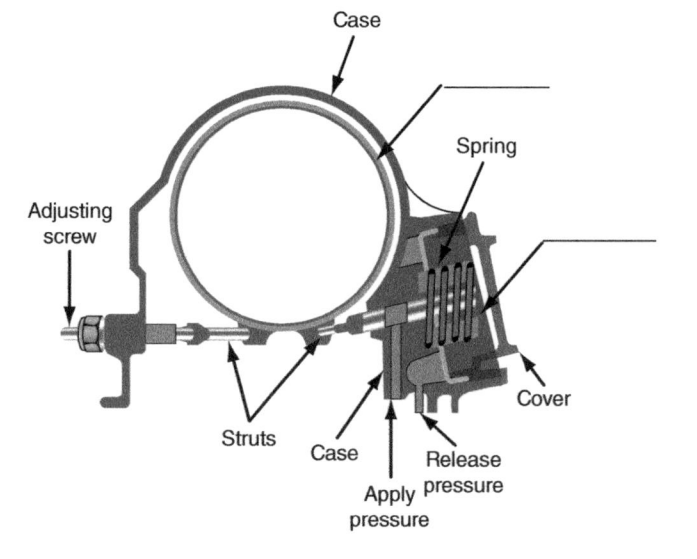

Case

Spring

Adjusting
screw

Struts   Case   Release
pressure

Apply
pressure

Cover

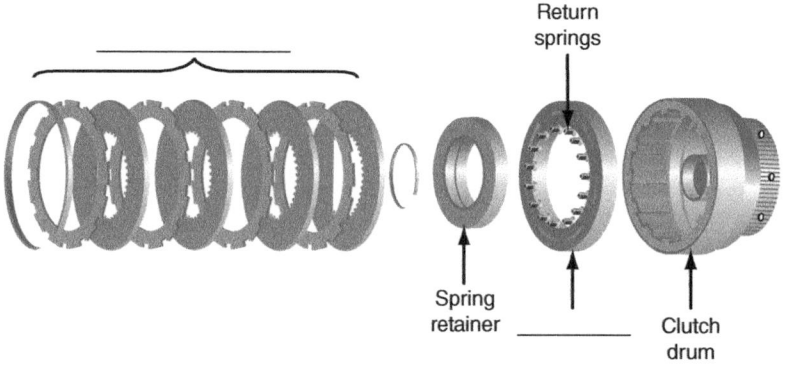

Return
springs

Spring
retainer   _____   Clutch
drum

A complete automatic transmission system for a front wheel drive vehicle (Honda) is shown below. Name the numbered parts.

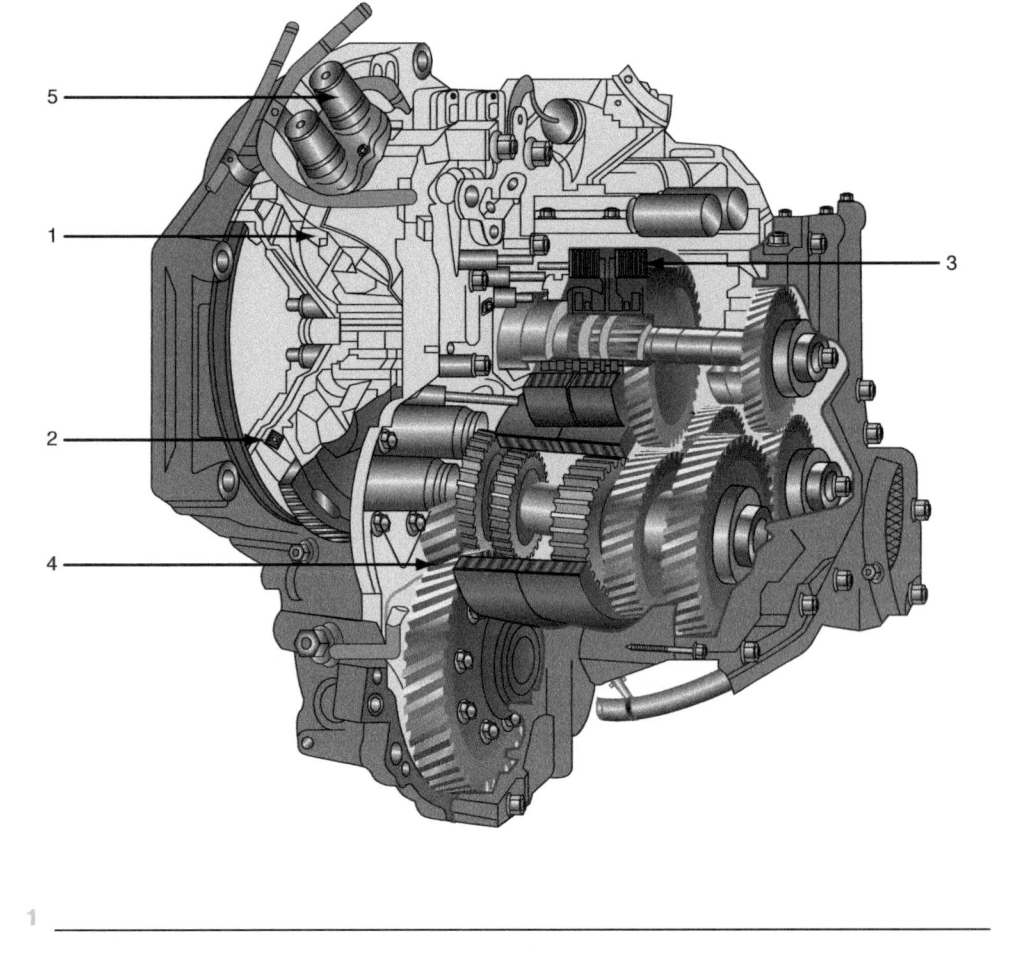

1 _____

2 _____

3 _____

4 _____

5 _____

State the function of the:

a   Accumulator

_____

_____

b   Lock-up clutch

_____

_____

OIL COOLER

A typical oil cooler system is shown.

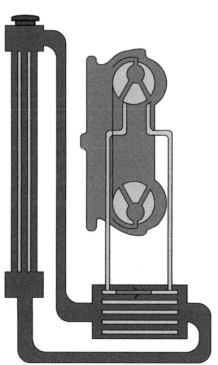

1 Label the drawing (add arrows to indicate flow).
2 State the purpose of the system.
3 Describe briefly how it operates.

_____

_____

_____

An oil cooler is very often a standard fitment on many automatic transmissions, it can however be fitted as a modification. Give reasons why an oil cooler may need to be added to an automatic transmission:

_____

GEAR SELECTION (AUTOMATIC TRANSMISSION)

A selector lever and quadrant showing the various operating positions for a four-speed automatic transmission are shown below.

State the purpose of each gear-selector position and button 'S' shown on the diagram on page 221.

P _____

_____

R _____

N _____

D _____

_____

3/2/1 _____

_____

_____

S _____

_____

Many modern automatic transmissions have a 'sequential' operating mode. Briefly describe what this means.

_____

_____

_____

Starter inhibitor

For safety reasons the starter can only be operated in selector positions P and N. This is achieved by an inhibitor switch on the gearbox which is operated by the manual selector mechanism. The four-terminal switch is operative in two positions for starter, or reverse-light operation.

Label the diagram below.

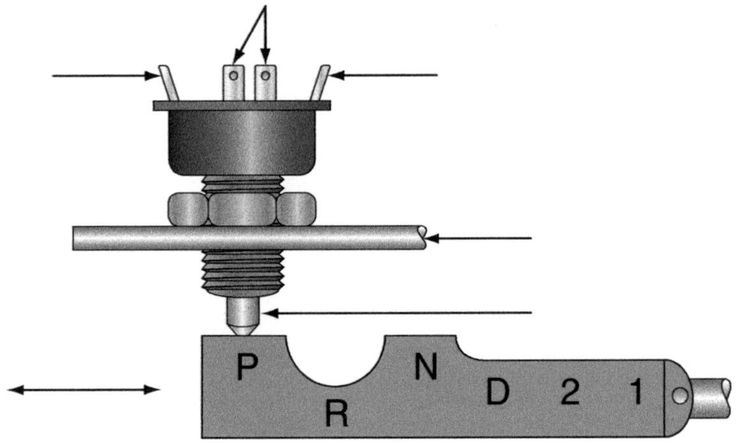

With an electrical shift, a multi-function switch informs the ECU of the selector lever position, this prevents the starter motor operating when the transmission is in gear and also controls the reversing lights.

State the precautions needed when towing a vehicle with automatic transmission when:

1 Engine is defective.

_____

2 Gearbox is defective.

_____

ELECTRONIC CONTROL

An electronically controlled automatic gearbox is basically the same as a hydraulically controlled automatic gearbox in that the main components are the torque converter, epicyclic gear set, multi-plate clutches and hydraulic servos.

In addition to these main components, electronic shifting entails the use of:

1 ECU (also known as) _____

2 _____

3 _____

Complete the simple block diagram for such a system and describe briefly how automatic gear changing is achieved.

SENSORS

1. ...vehicle speed............................ →

2. .... →

3. .... →

4. .... →

5. .... →

6. .... →

7. .... →

Operation

_____

_____

_____

_____

_____

Electronic control makes mode selection and operation very much easier and effective.

The layout of an electronic control system for an automatic transmission is shown below.

Name the lettered items:

a _____

b _____

c _____

d _____

e _____

f _____

g _____

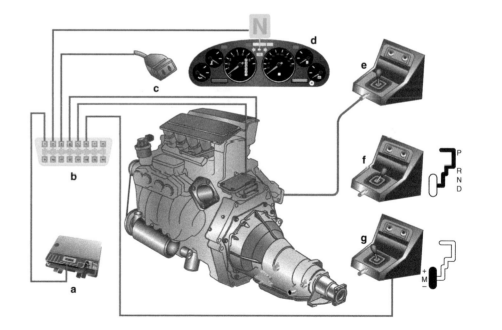

CONTINUOUSLY VARIABLE AUTOMATIC TRANSMISSION (CVT)

Continuously variable or stepless transmission is an alternative to the fixed ratio transmission. Unlike conventional automatic transmission, the gear ratios are varied in a smooth, stepless progression to suit driving condition (speed and load).

One transmission arrangement contains two basic elements:

1 **A planetary gear set integrated with two wet multi-plate clutches.**

2 **A belt and pulley system.**

The components named in (1) provide take up from rest and drive to the pulley system.

State the function of (2):

_____

_____

A front engine front wheel drive transaxle (Honda) is shown. Complete the labelling on the drawing.

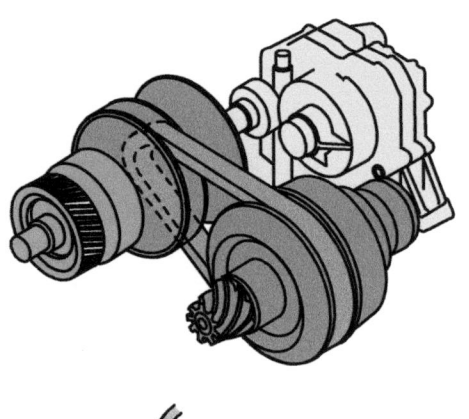

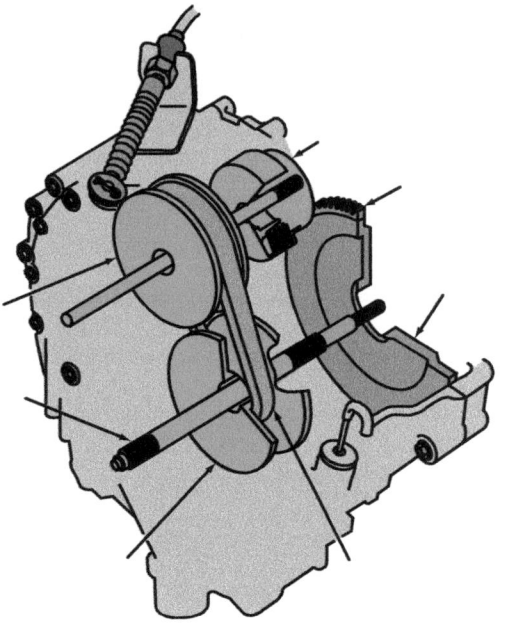

Honda's CVT

This system does not require a torque converter or a conventional friction clutch. Why is this?

_____

_____

CVTs use pulleys that change size and are connected by a belt.

List two vehicles using this (or similar) types of transmission:

Make _____ Model _____ Engine size _____

Make _____ Model _____ Engine size _____

The diagram below shows a typical layout and operating system of a CVT. Complete the labelling of this diagram.

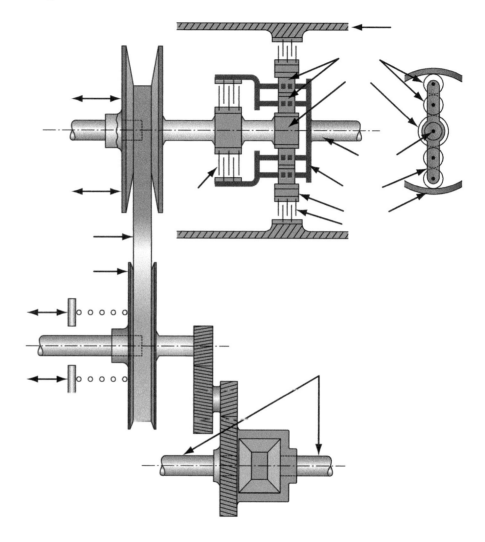

Operation

The forward clutch is gradually engaged by a hydraulic servo, the oil being supplied by an engine-driven oil pump via a control unit.

With the forward clutch clamped, the sun wheel and planet carrier are locked together making a fixed drive to the primary driving pulley.

Describe the action of the gear set in reverse.

_____

_____

_____

CONTROL

A hydraulic valve-control box determines the oil pressures which are applied to each part of the system: first to engage the appropriate clutch and then to select the optimum gearing ratio.

Driver inputs to the control box are transmitted mechanically via a flexible cable connected to the selector lever and by another cable which senses engine load from the throttle position.

Engine and road speed are measured by two Pitot tubes operating in rotating centrifugal chambers at appropriate points.

What is a Pitot tube?

_____

_____

Describe, with the aid of diagrams, the action of the belt and pulley system:

Low gear	High gear

_____

_____

_____

_____

Driving pulley width is controlled by: _____

Secondary or driven pulley width is controlled by:

_____

_____

From what material is the belt made?

_____

The basic layout and power flow for a STEPLESS transmission currently used by HONDA is shown below.

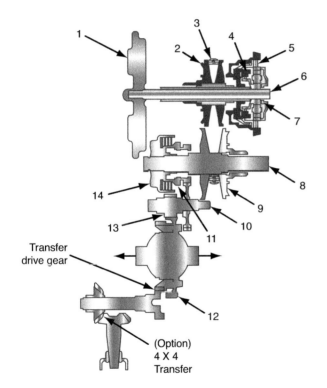

Transfer
drive gear

(Option)
4 X 4
Transfer

1. Flywheel
2. Drive pulley
3. Steel belt
4. Forward clutch
5. Reverse brake
6. Input shaft
7. Sun gear
8. Drive pulley shaft
9. Driven pulley
10. Secondary drive gear
11. Parking gear
12. Final driven gear
13. Final drive gear
14. Start clutch

Use arrows to indicate on the drawing the output for the front wheel drive shafts.

State the advantages of the continuously variable automatic transmission over the fixed ratio automatic transmission.

_____

_____

Operation

Basically with the forward clutch engaged (pressurised), drive from the engine is transmitted to the drive pulley. The belt transmits drive to the driven pulley and with the start clutch engaged drive is transmitted to the secondary drive gear and final drive.

The effective pulley ratio changes automatically with engine speed due to hydraulic pressure variation acting on the movable faces of the pulleys.

Describe the procedure for checking the fluid level in automatic transmission systems.

_____

_____

_____

_____

In addition to checking the fluid level, the condition of the fluid should be examined. Why is this?

_____

_____

_____

SEMI-AUTOMATIC TRANSMISSION

An alternative to a fully automatic transmission is a manual gearbox with robotised gearshift and clutch operation. This is a semi-automatic transmission with two pedal (no clutch pedal) control.

The layout below shows the main control features of this system.

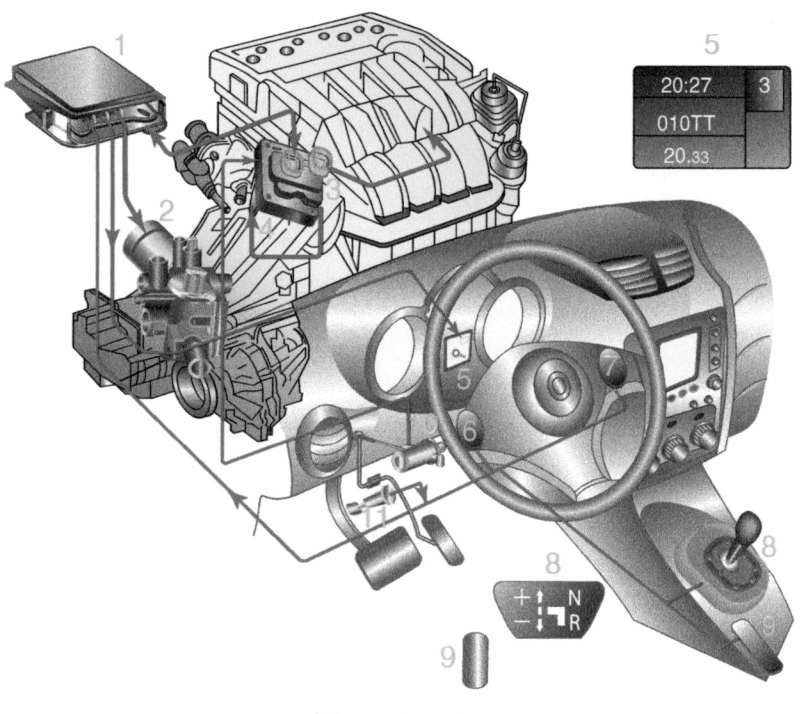

Selespeed gearbox

1 Electronic gearbox management unit
2 Clutch/gear shift actuator unit
3 Electronically controlled throttle
4 Electronic engine management unit
5 Display on fascia
6 'Down' control for shifting down through gears
7 'Up' control for shifting up through gears
8 Lever for sequential gear selection and display

9 'City' control for automatically activating gear changes
10 Electronic accelerator potentiometer
11 Switch on brake pedal

Study the illustration and complete the electrical circuitry to the schematic layout below.

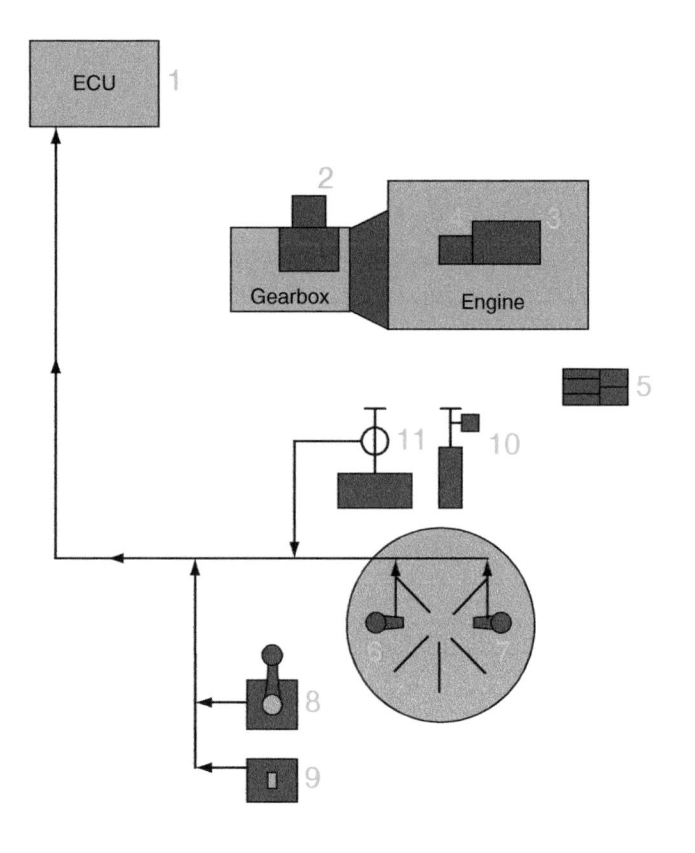

On receiving a signal from the gear lever (or steering wheel button) the ECU will energise solenoid valves in the clutch/gear shift actuator unit. This controls pressure in hydraulic servos to move the gear engagement, release and selection levers; and to operate the clutch. In 'city mode' fully automatic gear shifting is in operation.

The control of automatic transmissions is almost exclusively achieved through the use of electronically operated hydraulic systems. Actuation of the clutches is primarily by hydraulics. The electronics are responsible for the gear selection and adapting the pressures depending upon the torque flow.

State the advantages of using this type of system:

1 _____

2 _____

3 _____

4 _____

5 _____

A number of sensors are used to detect the engine load and speed; transmission output shaft speed; selector lever position; as well as the positions of the programme selector and kick-down switch. The electronic control unit (ECU) processes the information against a predefined programme, the results of which are used to control the variables of the transmission (which gears and when they are selected).

The link between the electronic and hydraulic circuits is by means of electro-hydraulic converter elements. The clutches are activated by traditional solenoid valves. The pressure levels at the friction surface is precisely controlled by analogue or digital pressure regulators.

Shift-point control

The selector lever allows for manual input as well as different shift programmes, which can include maximum fuel economy and maximum performance. The system refers to the rotational speeds of the transmission output shaft and the engine before the appropriate solenoid valves are triggered.

How do intelligent shift programmes improve driveability?

_____

_____

_____

Converter lock up

Torque converter slip is eliminated by employing a mechanical lock up clutch, which helps to improve transmission efficiency. The converter lock up clutch is activated under certain conditions that are determined depending on engine load and transmission output speed.

Control of shift quality

Engine load and speed are used to determine the accuracy at which the pressure at the friction elements are adjusted to the level of the torque being transmitted. This has a decisive effect on shift quality, the pressure of which is regulated by a pressure regulator. By briefly reducing engine torque for the duration of the shift, which is achieved by retarding the ignition timing, shifting comfort is then further enhanced. This also extends the service life of components by reducing friction losses at the clutches.

Safety circuits

To prevent damage to the transmission, due to driver error, the system has special monitoring circuits. The system reverts to a backup mode in response to any electrical malfunctions.

Final-control elements

The final interface between electric and hydraulic circuits is achieved with the use of solenoid valves and pressure regulators.

TESTING AND FAULT DIAGNOSIS

Torque converter

One method of testing a torque converter coupling is to carry out a stall test on the vehicle. Outline this procedure, including any safety measures adopted during the test:

1 _____

2 _____

3 _____

4 _____

5 _____

_____

_____

For satisfactory operation the engine stall speed should be approximately _____ revs/min. Indicate the possible faults for the results given below.

Engine speed 300 revs below specified stall speed:

_____

Engine speed 600 revs below specified stall speed:

_____

Stall speed too high:

_____

_____

A road or chassis dynamometer test is another way in which torque converter faults are diagnosed; the road test will also confirm faults suspected during a stall test.

State how the road test would indicate:

1 Slipping stator

_____

_____

2 Seized or locked stator

_____

_____

If the converter pressure is low, 'cavitation' will occur, resulting in slippage, noise and vibration. What are the effects if converter pressure is too high?

_____

_____

With the faults at 1 and 2 above it is necessary to replace the converter as it is normally a sealed unit.

> ⚡ The fluid may be extremely hot; it is therefore necessary to avoid direct skin contact with it or preferably allow a period of time for it to cool down.

The fluid coupling requires very little maintenance apart from topping up with fluid; however, leakage could result from a worn seal and worn bearings could allow the faces of the impeller and rotor to come into contact, giving noisy operation.

How can the system be protected during use or repair against the following hazards?

1 Overheating

_____

_____

2 Aeration and foaming

_____

3 Contamination of fluid

_____

4 Ingress of dirt and moisture

_____

5 Fluid leakage

_____

_____

6 Mechanical damage

_____

Routine preventive maintenance will improve reliability and efficiency, and maximise the life of a transmission system. List the preventive maintenance checks and tasks associated with automatic transmission:

Check:

- _____
- _____
- _____
- _____
- _____
- _____
- _____

- _____
- _____
- _____
- _____
- _____

List the general rules for efficiency and any special precautions to be observed when carrying out maintenance on the transmission:

- _____
- _____
- _____
- _____
- _____
- _____
- _____
- _____
- _____

Oil can be hot on transmissions that have just been tested – take care.

DIAGNOSTICS: AUTOMATIC TRANSMISSION: SYMPTOMS, FAULTS AND CAUSES

State a likely cause for each symptom/system fault listed in the table on page 232. Each cause will suggest any corrective action required.

Symptoms	Faults	Probable causes
Loss of drive; flare up on changes	Worn multi-plate clutch	
Incorrect gear selection; starter operates in other than P or N	Worn or incorrectly adjusted selector linkage	
Incorrect gear-change interval	Broken throttle signalling device (cable or vac pipe)	
Incorrect gear-change interval	Incorrectly adjusted kickdown cable	
Harsh engagement of gears, excessive creeping	High engine idling speed	
Slipping clutches	Low line pressure	
Bumpy gear shifts	High line pressure	
Slip in one gear	Oil leakage	
Park will not hold car	Broken parking pawl	

Electronically controlled automatic transmission (ECAT)

List the common electrical faults which could occur with an ECAT system:

- _____
- _____
- _____
- _____
- _____
- _____
- _____

Multiple choice questions

Choose the correct answer from a), b) or c) and place a tick [✓] after your answer.

1 **The purpose of an automatic transmission stall test is to check the:**

a) The brake bands and hydraulic servos [ ]

b) Torque converter and transmission clutch slip [ ]

c) Lock-up clutch and brake bands [ ]

2 **One possible cause for automatic transmission fluid being discoloured and having a burnt smell is:**

a) Brake bands seized [ ]

b) Kickdown cable too tight [ ]

c) Slipping clutches [ ]

3 **An epicyclic gear train consists of a:**

a) Sun gear, planet gears and annulus [ ]

b) Sun gear, planet gear and pump [ ]

c) Sun gear, moon gears and annulus [ ]

SECTION 5

Drive line

USE THIS SPACE FOR LEARNER NOTES

_____

_____

_____

_____

_____

_____

_____

_____

Learning objectives

After studying this section you should be able to:

- Be able to explain the construction of transmission and drive line.
- Know how the electrical and electronic systems operate and interact with mechanical systems in transmissions and drive line.
- Be able to explain the types of faults and their causes in transmission and drive line systems.
- Know how to systematically diagnose faults.

Key terms

Differential Divides the transmitted torque equally to each output shaft and allows each to rotate at different rates.

Crown wheel Large gear wheel used in a differential.

Pinion gear A gear which meshes with, and directly drives the crown wheel.

Three quarter floating Type of bearing arrangement used in vehicle hub assemblies.

WWW www.zf.com

www.howstuffworks.com/differential.htm

DRIVE LINE SHAFTS AND HUBS

What is the meaning of the term 'dead axle'?

_____

Complete the sketch below to show a dead axle:

 **TIP** Always follow manufacturer's instructions and use data for tightening when replacing drive line components.

Name two vehicles that have dead axles:

Make   Model

Make   Model

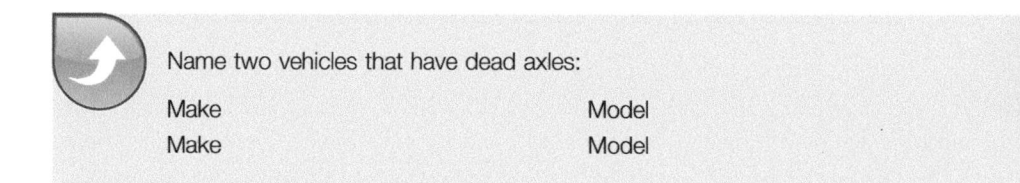

Bearing arrangements in dead axle hubs are similar to those already described; however they are not classed as semi, three-quarter or fully floating. Why is this?

_____

_____

DIFFERENTIAL

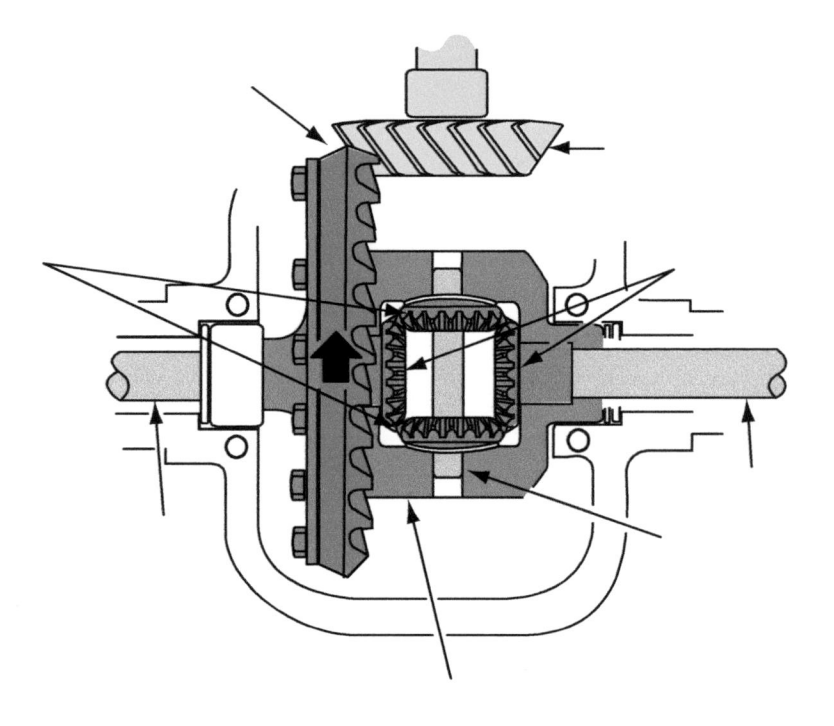

The components of a typical final drive (differential) unit

Label the diagram above to show the final-drive gears for a front-wheel-drive transverse engine car and give a typical gear ratio for this arrangement.

Vehicle make   Model

Ratio

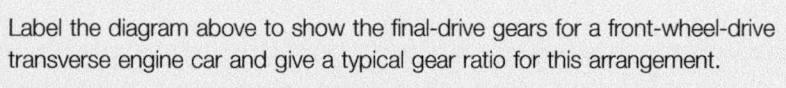

The purpose of the differential is to transmit equal driving torques to the half shafts while allowing the shafts to rotate at different speeds when the vehicle is travelling in other than a straight line.

 Observe the action of a differential by rotating the sun wheels at different speeds to each other and describe the action of the planet wheels

The main components of a differential are:

_____

Investigation

1 Remove the final-drive assembly (either bevel gear or worm and wheel) from an axle casing.

Describe the method of pre-loading the bearings.

_____

_____

_____

2 Pinion bearings:

_____

Side bearings:

_____

3 State the backlash setting and the pre-load setting for the assembly.

Backlash _____ Pre-load (pinion) _____

(side bearings) _____

4 Assemble the unit and describe briefly the procedure for adjusting the final drive gears to the manufacturer's specifications.

A _____

_____

B _____

C _____

D _____

_____

E _____

LIMITED-SLIP DIFFERENTIAL (LSD)

Sports and racing cars are vehicles with high power-to-weight ratios and as such can, even on good surfaces, cause a driving wheel to spin, say, during rapid acceleration or during hard cornering. Most of the higher quality or top-of the range models incorporate an LSD.

The limited-slip device is incorporated in the differential, which automatically applies a brake to the spinning half shaft, thereby maintaining a torque in the other half shaft.

Little torque is required to drive a spinning or slipping road wheel and a normal type differential would transmit the same torque to the non-spinning wheel; thus traction is lost. Limited slip traction control and torque sensing devices all produce 'frictional torque' in the drive to a spinning wheel.

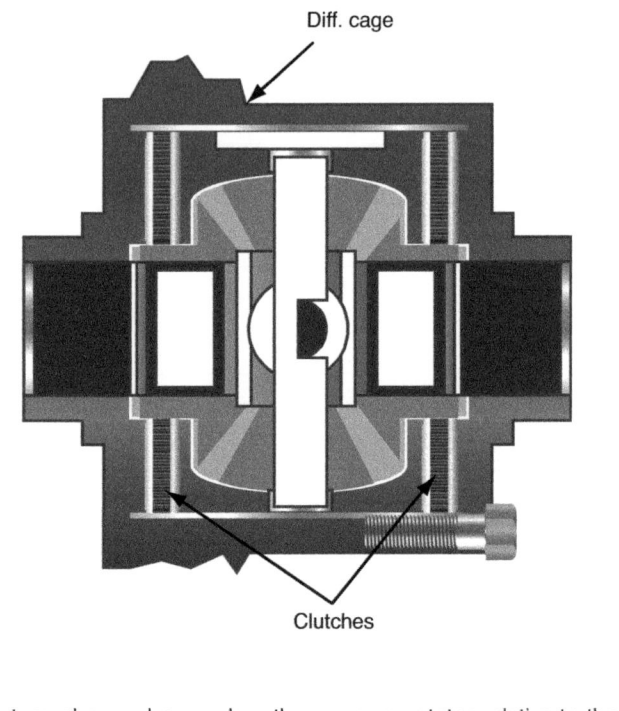

Diff. cage

Clutches

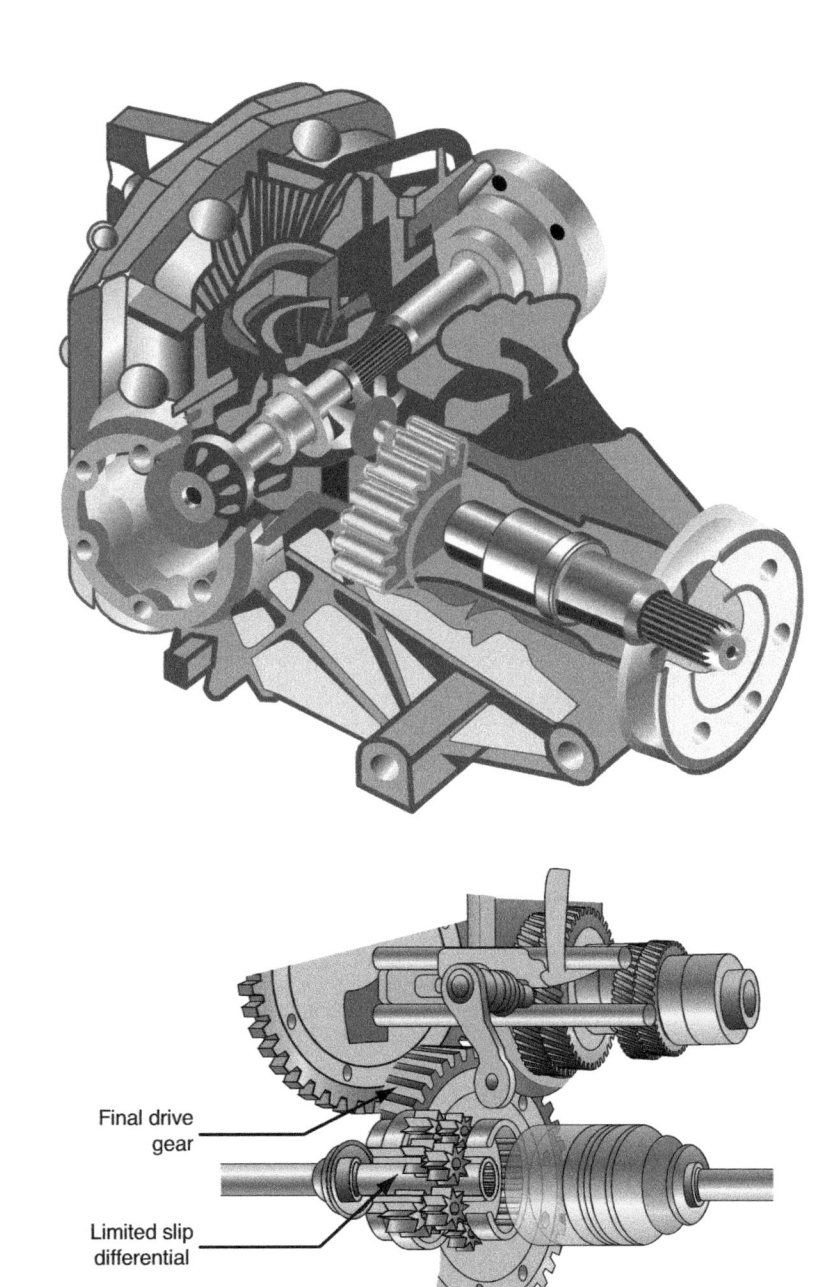

Final drive gear

Limited slip differential

Limited slip differential

In the friction clutch type shown above, when the sun gear rotates relative to the differential cage (during wheel spin), the spring loaded clutch slips generating frictional torque. Owing to the 'torque balancing' action of a differential the same torque is transmitted to the non-spinning wheel.

Describe briefly how to check a limited-slip differential for operation and state any safety precautions to be observed when working on a vehicle fitted with a limited slip differential.

_____

_____

_____

INTER AXLE (THIRD) DIFFERENTIAL

If the drive to the two final-drive pinions or worm gears is transmitted through a third differential, speed variation between the two sets of final-drive gears can occur and the driving torque is shared equally between the two axles.

The simplified drawing below shows a worm and wheel tandem drive incorporating a third differential. Describe the operation of this system.

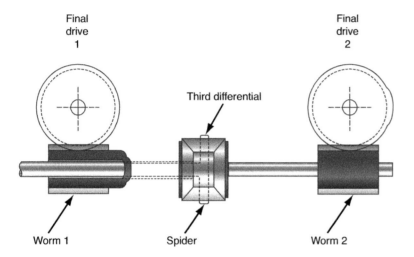

Final drive 1

Final drive 2

Third differential

Worm 1

Spider

Worm 2

_____

_____

_____

Why is it necessary to have a differential lock on the third differential?

_____

_____

DRIVE LINE, SHAFTS AND HUBS: SYMPTOMS, FAULTS AND CAUSES

State a likely cause for each symptom/system fault listed below. Each cause will suggest any corrective action required.

Symptoms	Faults	Probable causes
Backlash on take up of drive and vibration	Worn universal joint trunnions	
Lubricant leakage	Split rubber gaiters	
Noise from hub; excessive hub end float; overheating of hub assembly	Worn hub bearings	
Transmission vibration	Worn centre bearing	
Oil leakage	Worn sliding joint oil seal	
Knocking noise when turned on lock; vibration	Worn constant velocity joint	
Knocking on take up of drive	Worn driveshaft splines	
Transmission vibration and noise	Loose shaft flange bolts	
Transmission vibration and noise	Sheared rubber (doughnut type universal joint)	

⚡ Dispose of waste correctly and recycle where appropriate.

Drive line shafts and hubs: protection during use

How can drive line systems be protected against the ingress of moisture and dirt during use and repair?

- _____
- _____
- _____
- _____

Maintenance

Routine maintenance is essential in order to ensure reliability, maintain efficiency, prolong service life and ensure vehicle safety. List the major maintenance points:

1 _____

2 _____

3 _____

4 _____

5 _____

6 _____

7 _____

8 _____

9 _____

Describe any special tools (and their care) necessary to carry out routine maintenance and adjustments:

- _____
- _____
- _____
- _____
- _____
- _____

List the general rules/precautions to be observed when carrying out routine maintenance adjustments, removal and replacement relative to the drive line, shafts and hubs:

- _____
- _____
- _____
- _____
- _____
- _____
- _____
- _____

Describe the procedure for removing and replacing a CV joint book using the images below.

Step 1 _____
_____
_____
_____

Step 2 _____
_____
_____

Step 3 _____
_____

Step 4 _____
_____
_____
_____

Step 5 _____
_____

Step 6 _____
_____

Step 7 _____
_____
_____

Step 8 _____
_____

239

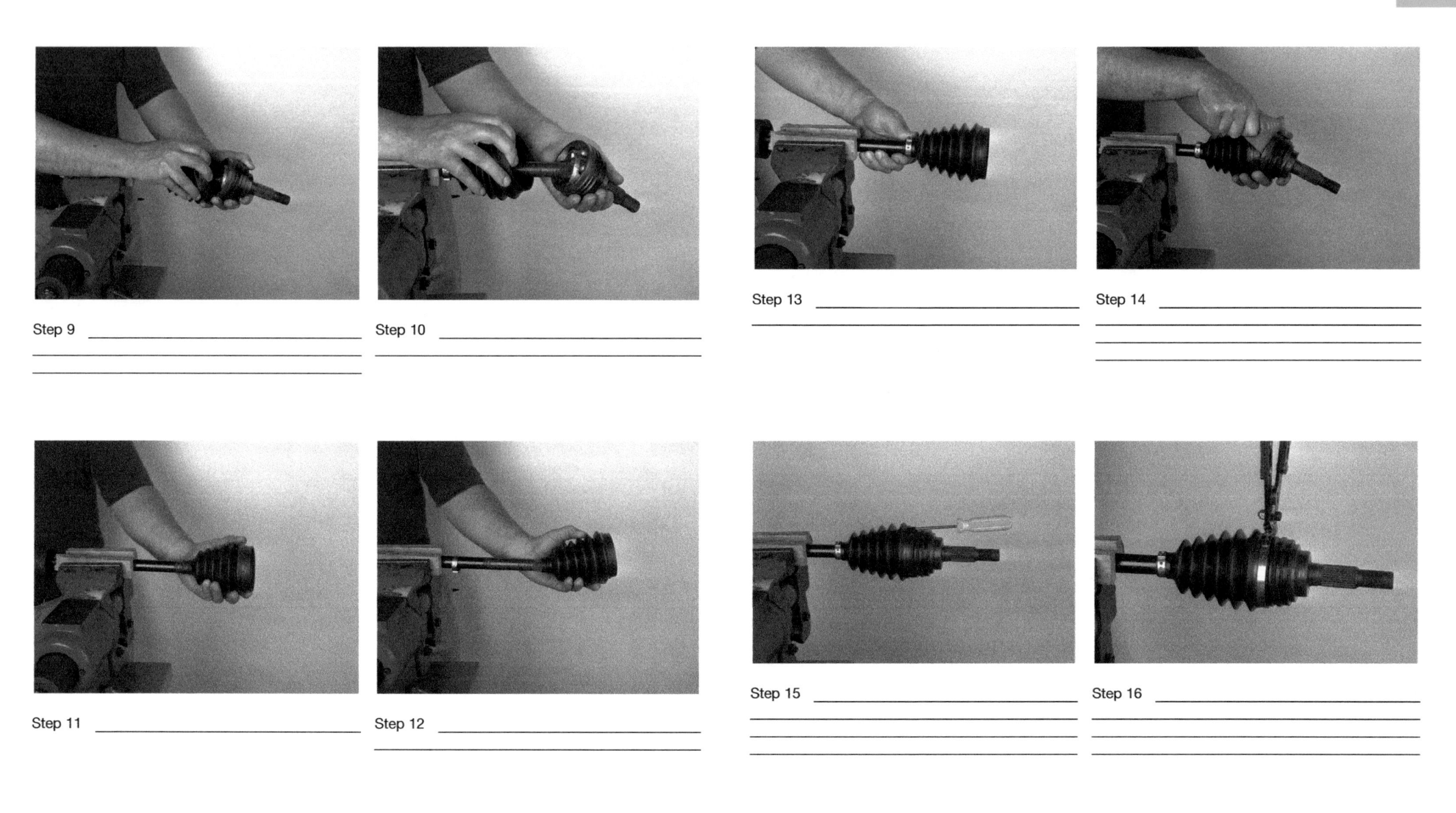

Step 9 _____
_____
_____

Step 10 _____
_____

Step 11 _____

Step 12 _____
_____

Step 13 _____
_____

Step 14 _____
_____
_____

Step 15 _____
_____
_____
_____

Step 16 _____
_____
_____

FINAL DRIVE AND DIFFERENTIAL DIAGNOSTICS: SYMPTOMS, FAULTS AND CAUSES

State a likely cause for each symptom/system fault listed below. Each cause will suggest any corrective action required.

Symptoms	Faults	Probable causes
Oil spray on underside of floor and chassis; evidence on ground	Oil leakage	_____ _____
Transmission noise	Bearing failure	_____ _____ _____
Noise on drive or over run; backlash on take up of drive	Excessive end float in pinion bearings	_____ _____
Transmission knock or tick; vibration	Damaged or broken final-drive gears	_____ _____ _____ _____
Transmission noise	Worn differential gears	_____ _____ _____ _____
Excessive slip; loss of traction under arduous conditions	Faulty VC or worn clutch pack	_____ _____ _____
Knocking on take up of drive	Spline damage	_____ _____ _____
Transmission noise; backlash	Incorrect meshing of final-drive gears	_____ _____ _____

Multiple choice questions

Choose the correct answer from a), b) or c) and place a tick [✓] after your answer.

1 **A non-driven type of axle is known as a:**

   a) Live axle [ ]

   b) Dead axle [ ]

   c) Third axle [ ]

2 **The purpose of a differential is to allow:**

   a) The outer wheel to turn faster [ ]

   b) Both wheels to turn at equal speeds [ ]

   c) The wheels to rotate at different speeds [ ]

3 **Which one of the following, is the correct procedure when carrying out a routine vehicle inspection:**

   a) Drive shaft gaiters should be checked for splits [ ]

   b) The CV should be repacked with grease [ ]

   c) All driveshaft gaiters should be replaced [ ]

SECTION 6

Electronic components

USE THIS SPACE FOR LEARNER NOTES

_____

_____

_____

_____

_____

_____

Learning objectives

After studying this section you should be able to:

- Understand the basic functions of an electronic control unit (ECU).
- Be able to explain the purpose and function of actuators and sensors.
- Understand the meaning of 'mechatronics'.
- Describe how multiplexing works.
- Know the basic operating principles of fibre optics.
- Know how to systematically diagnose faults.

WWW   www.picoscope.com

www.autoshop101.com

Key terms

Sensor   An electrical component that changes its resistance or electrical current in response to the changes in a vehicle's characteristics e.g speed sensor.

Actuator   An electrical component that moves in response to a command e.g. a solenoid.

Multiplexing   A system which uses only two wires (CAN) for a number of ECUs to send and receive information along.

Fibre optics   Data signals in the form of light are transmitted along thin glass or plastic fibres.

Electronic control unit (ECU)

A number of sensors are used on the modern electronically controlled transmission system. The information from these is received by the ECU in either digital or analogue formats, the latter needing to be converted to digital signals by an analogue-to-digital converter inside the ECU. The ECU uses this information to decide which actuator, or actuators, it needs to operate and for how long.

SENSORS AND ACTUATORS

Temperature sensor

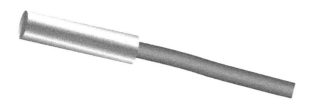

An NTC temperature sensor

It is important that the transmission oil temperature is maintained within the manufacturer's specified tolerances. This is monitored by a temperature sensor, the speed and accuracy of which is imperative. Typical operating temperature ranges for this type of sensor are –55°C to +180°C.

The type of sensor used is an NTC. This is an abbreviation of: _____

Testing of an NTC sensor is as follows:

_____

_____

It is possible with most modern systems to check the operation of the NTC with the use of a scan tool.

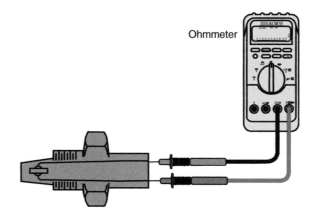

Ohmmeter

A typical NTC being tested

As with any system checks and diagnostics, refer to the manufacturer's technical data.

Complete the following graph showing the line trace of how the resistance changes due to the increase of temperature.

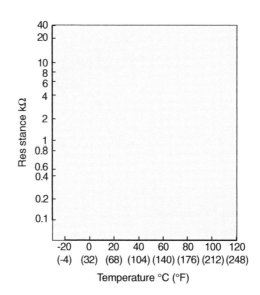

Solenoid valves

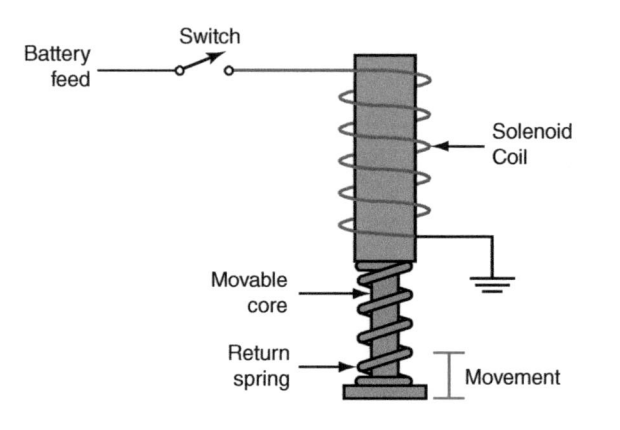

Solenoids are used in a number of vehicle applications. In transmission systems they are generally

used to operate valves. These valves are used to: _____

_____

Solenoids are electromechanical magnets, consisting of insulated copper wire windings, wound around a moveable centre core. This centre core is what operates as the valve.

General test of these solenoids can be carried out using a digital multi meter (DMM) set to read

_____ and thereby checking the _____

The readings can tell us if there is an _____

Speed sensors

Label the diagram below using the following terms:

vehicle speed sensor
o-ring (two labels required)
speedometer pinion gear

speedometer pinion gear adapted
sensor mounting bolt
electrical connector

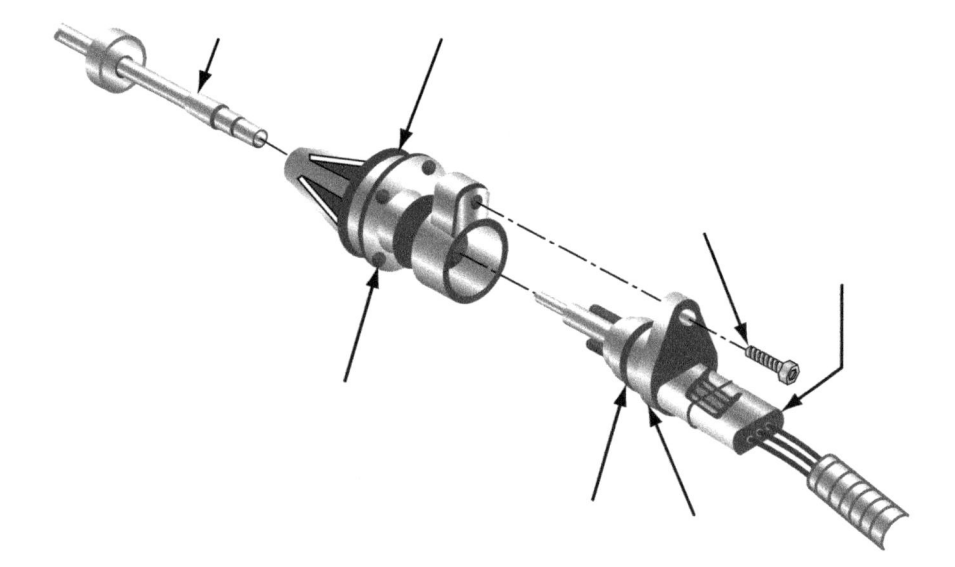

The input and output speed of the gearbox is generally scanned by a digital Hall type speed sensor, however, analogue inductive type sensors maybe used by some vehicle manufacturers. These signals are needed by the ECU to determine the speed difference between the input and output, in order to select the correct gear, based upon other information received from other vehicle systems via the CAN. The sensors monitor the rotational speeds of shafts. These signals can also be used to determine if there is clutch slip within the transmission system. This is

achieved by _____

_____

The type of signal produced by a Hall type speed sensor is a _____

 In small groups discuss the types of signals produced by an inductive and Hall type sensor. Draw a diagram showing each of the signals for these sensors. Include and explain the amplitude and frequency of these signals. Where possible, find an appropriate vehicle to connect an oscilloscope to and check these signal waveforms.

Pressure sensors

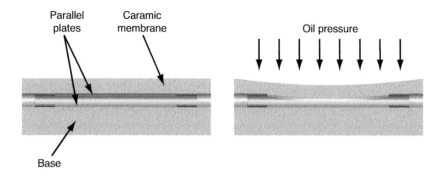

Capacitance type pressure sensor. As the membrane is deformed the distance between the two plates changes and so does the capacitance voltage

To ensure good performance inside an automatic transmission the hydraulic pressure needs to be correctly maintained under all performance conditions. As the temperature of the oil increases, the viscosity of the oil reduces, which can in turn affect the pressure in the system. The type of sensor

used is similar to that of the engine pressure sensor: _____

_____

The sensor technology used is a: _____

_____

_____

Drive train torque sensor

These sensors are used for monitoring the efficiency of the power train, which helps to prevent damage to the transmission. They operate by using non-magneto-resistive sensor technology, to either measure the stress on a shaft or the relative angular movement on a shaft.

Travel sensors

On some transmission systems, such as the DSG, travel sensors are used to detect that a gear is engaged or disengaged. These use a permanent magnet fitted in each selector fork to allow for the detection of gear selection.

Mechatronics

A mechatronics unit is used in a number of modern electronically controlled transmission systems (as well as other vehicle applications). Mechatronics is the terminology given to a self-contained unit which incorporates electronic and mechanical/hydraulic engineering components and systems which interact with each other. They can contain the hydraulic valve body, on/off solenoids, transmission ECU, speed, position, temperature and pressure sensors. Some versions also incorporate shift rail actuators.

The main advantages of using a mechatronics unit are: _____

_____

Selector lever

These types of levers mimic a traditional gear lever in appearance. They are not mechanically connected to the transmission system in a means or form as they communicate with the ECU electronically. These communications are by means of Hall effect sensors situated in the base of the selector lever.

A Hall effect sensor produces a: _____

The ideal piece of diagnostic equipment to check this signal is an: _____

Multiplexing

The typical modern transmission system will integrate with the other vehicle systems via the transmission control unit (TCU), such as with the engine control unit. The communication between the systems is achieved using the CAN Bus system.

The transmission system may need to communicate, via the CAN Bus, with the engine management system. The information which the transmission ECU will need to use, in order to select the optimum gear for a set period of time, will be based on the engine load and speed.

The advantages of having integrated control systems are: _____

_____

_____

_____

_____

_____

Fibre optics

Fibre optics are becoming commonly used on vehicle sensing systems, especially with the increase in vehicle safety systems. They can be used in a number of traction control systems.

The basic operating principle of fibre optics is a cold beam of pulsed light is passed through a plastic or glass fibre optic cable, which is as thin as a human hair. Light does not bend and they bounce off the reflective inner walls of the fibre optic cables, transmitting data quickly, over large distances.

A transmitter is at one end of the fibre optic cable, which transmits the data in light pulses along this cable. The information enters at the other end into a receiver, where it is decoded and used by the ECU to activate actuators.

Multiple choice questions

Choose the correct answer from a), b) or c) and place a tick [✓] after your answer.

1 **In a multiplexed vehicle wiring system:**

   a) The supply voltage is always very high [ ]

   b) More wire is required than conventional systems [ ]

   c) The data bus carries signals to and from ECUs [ ]

2 **An NTC is a:**

   a) Negative temperature coefficient sensor [ ]

   b) Negative temperature coefficient actuator [ ]

   c) Negative temperature computer actuator [ ]

3 **Which one of the following statements is correct?**

   a) Actuators send electrical signals to the electronic control unit [ ]

   b) Fibre optic cables transmit light signals along glass or plastic fibres [ ]

   c) Sensors receive electrical signals from the electrical control unit [ ]

PART 6
ELECTRICAL SYSTEMS

USE THIS SPACE FOR LEARNER NOTES

_____

_____

_____

_____

_____

_____

_____

SECTION 1
Electronic principles, lighting and infotainment systems 250

1 Electronic principles 251
2 Electronic systems – abbreviations and symbols 251
3 Waveforms 252
4 Resistance 253
5 Capacitance 254
6 Switches and relays 255
7 Sensors 257
8 Transistors 258
9 Multiplex 260
10 Lighting circuits and regulations 262
11 Diagnostics: lighting – symptoms, faults and causes 266
12 Infotainment systems 267
13 Multiple choice questions 272

SECTION 2
Starting and charging systems 273

1 Types of starter motor 274
2 Permanent magnet starters 275
3 Testing/diagnosing starter motor faults 276
4 Charging systems 278
5 Testing/diagnosing alternator faults 279
6 Liquid cooled alternator 281
7 Smart charging systems 282
8 Stop/start technology 282
9 Multiple choice questions 283

SECTION 3
Vehicle body electrical systems 284

1   Auxiliary electrical systems 285
2   Windscreen wiper systems 285
3   Electrically operated windows and mirrors 288
4   Screen heating 291
5   Power seats 292
6   Central locking systems 293
7   Heating, cooling and airconditioning 294
8   Multiple choice questions 303

SECTION 1

Electronic principles, lighting and infotainment systems

USE THIS SPACE FOR LEARNER NOTES

Learning objectives

After studying this section you should be able to:

- Explain the basic electronic principles.
- Identify common electronic symbols.
- Identify common electronic components.
- Describe multiplex systems.
- Identify lighting circuits and explain how they work.
- Describe different types of infotainment systems.
- Determine how to diagnose faults on electrical systems.

Key terms

Voltage The electrical potential difference between two points.

Current The flow of electric charge through an electrical conductor.

Resistance The opposition to the flow of electrical current.

CAN Controller area network.

MOST Media Orientated System Transportation.

Capacitor An electronic component that can store a charge for a period of time.

Transistor An electronic component that acts like a switch.

www.allaboutcircuits.com

www.mostcooperation.com

http://electronics.howstuffworks.com

www.physlink.com

ELECTRONIC PRINCIPLES

Use the words below to complete the following paragraph:

computers	heat	technology	outputs
process	inputs	systems	conventional
electrical	illuminate	sensors	
integrated	electronic	transmit	

Electrical _____ are usually concerned with using electricity to

_____ energy, for example _____ light bulbs use

_____ energy and convert it to _____ energy to make it

_____. _____ systems are more concerned with using

electricity to _____ information. This is done through _____, micro

components and _____ circuits to receive _____ and transmit

_____ to specific components using very similar _____ to modern

day _____.

Name two types of devices that would be considered:

a   Electronic

1   _____

2   _____

b   Electrical

1   _____

2   _____

ELECTRONIC SYSTEMS – ABBREVIATIONS AND SYMBOLS

When working with electronic systems there are a number of abbreviations that relate to them, one of the most common being the ECU which is the electronic control unit. Identify the following abbreviations relating to electronic systems:

LED _____

LCD _____

ROM _____

RAM _____

BCM _____

EOBD _____

CPU _____

CAN _____

PCB _____

IC _____

In order to understand electronic system wiring diagrams a number of symbols are used. Complote thc table on page 252 lu iuenlIfy some basic electronic symbols, and give a brief description of each component.

Component	Symbol	Description
Transistor		
Capacitor		
Diode		
Variable resistor		
Thermistor		
Relay		
Thyristor		
Light dependent resistor		
Light emitting diode		
Fuse		
Resistor		
And gate		
Nor gate		
Zener diode		

WAVEFORMS

Electronic circuits can be monitored and checked for serviceability by using an oscilloscope. This will produce a number of waveforms dependent on the item being checked. For example a Hall effect sensor would produce a square wave pattern. Identify the following waveforms that could be found when using an oscilloscope:

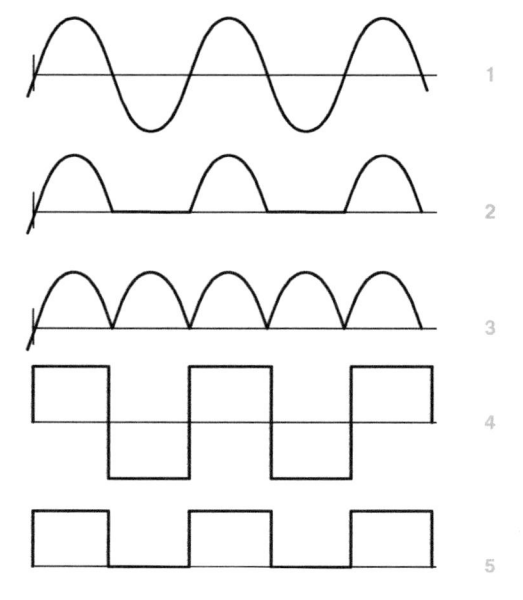

1 _____

2 _____

3 _____

4 _____

5 _____

CHECK Oscilloscopes give a visual interpretation of the electronic signals being delivered by components.

When using waveform patterns to diagnose faults it is useful to understand the various terms that are related to these patterns. Define the following terms:

Cycle: _____

Period: _____

Frequency: _____

Amplitude: _____

_____

Wavelength: _____

 Draw a sine wave pattern over a five second period that has 10 cycles per second. The wave should start at 0 volts and rises to +9 volts before dropping to −9 volts. Mark on the sine wave the amplitude and wavelength and calculate the frequency and period of the sine wave.

Being aware of the correct waveform patterns that would be expected by specific electronic components would allow quicker diagnosis of faults should there be an abnormality in the displayed pattern.

RESISTANCE

Resistance is described as the hindrance to the flow of electrical current. In electrical circuits the resistance can be affected by the length and size of the wires. This can be seen by the different sizes of cables used for the starter motor and the lighting circuits for example.

In electronic circuits, or when using electronic components, a resistor is fitted. The resistor is an electronic component that is used to control/limit the amount of voltage and current that flows

to particular components. For example, if an LED light requires around 2 volts and 0.02 amps to operate, if this was connected directly to a 12 volt battery the voltage would be too high and too much current would flow resulting in damage to the LED. The following shows a variety of resistors:

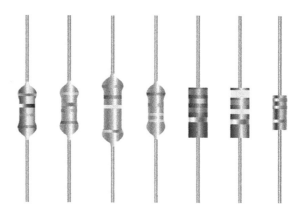

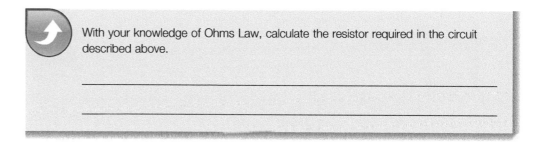 With your knowledge of Ohms Law, calculate the resistor required in the circuit described above.

_____

_____

Resistors come in a variety of values to fit individual circuits and circuit designers have to ensure the correct resistor is used to prevent damage to the circuit. To ensure the correct resistor is chosen for the circuit and components their value is indicated by a colour coded band system; the following table shows how resistors are coded:

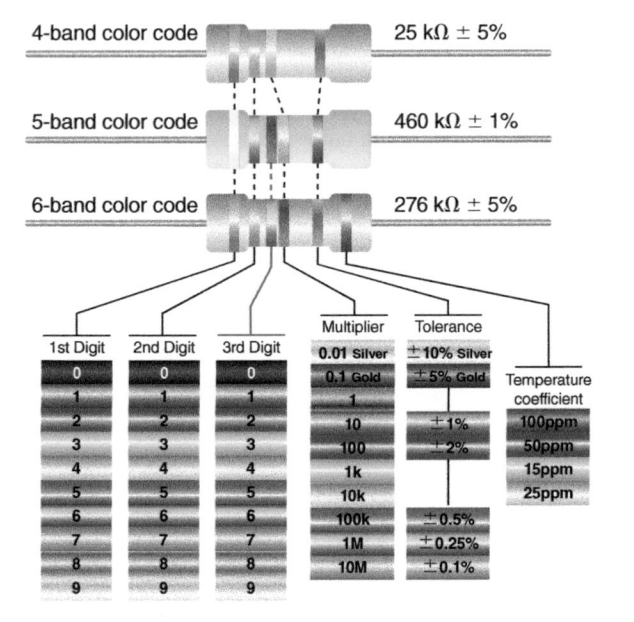

4-band color code — 25 kΩ ± 5%

5-band color code — 460 kΩ ± 1%

6-band color code — 276 kΩ ± 5%

1st Digit	2nd Digit	3rd Digit	Multiplier	Tolerance	Temperature coefficient
0	0	0	0.01 Silver	±10% Silver	
1	1	1	0.1 Gold	±5% Gold	
2	2	2	1		100ppm
3	3	3	10	±1%	50ppm
4	4	4	100	±2%	15ppm
5	5	5	1k		25ppm
6	6	6	10k		
7	7	7	100k	±0.5%	
8	8	8	1M	±0.25%	
9	9	9	10M	±0.1%	

Using the above diagram, identify the colour coding on the 600Ω resistor with a tolerance of + or − 5%.

Identify six different resistors and work out their values from their colour coding using the diagram above.

CAPACITANCE

A **capacitor** basically consists of two conductive plates separated by a dielectric (non-conducting substance). The dielectric can be substances such as glass, paper or air. A capacitor is charged up when a voltage is applied across the plates, once the potential difference across the plates is the same the capacitor will be fully charged.

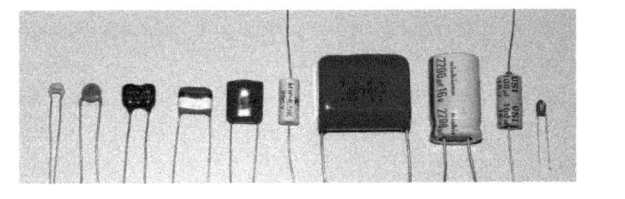

What is the function of a capacitor?

Capacitance is measured in farads (F). However, more often the capacitance is quite small in vehicle circuits and will be measured in micro farads or pico farads. The annotation for micro farads is µF using the Greek symbol µ for micro; a micro farad is one millionth of a farad. The annotation for pico farad is pF; a pico farad is one billionth of a farad.

The maximum working voltage of any capacitor is dependent on the thickness and type of dielectric used. The discharge rate of the capacitors also depends on the circuit it is connected to and its leakage rate. How would the capacitor be affected if it was subjected to a voltage that exceeded the maximum voltage?

Capacitors can hold high voltages so do not handle unless you are certain it has been discharged or you could discharge the capacitor and get a nasty shock.

SWITCHES AND RELAYS

A switch in a circuit is used to make or break the flow of current in the circuit. Depending on the application, switches can be simple on/off spring-loaded toggle types or more complicated multi-function switches such as the stalk type steering column arrangements.

In many circuits on a motor vehicle the manually operated switches energise relays which in turn carry the main current load for the circuit. The current required to energise the relays can be relatively small (0.3 amps) but once operated they allow a larger current to pass through them (20 amps). This allows smaller cables to be used to operate the relays which in turn reduce the overall weight of copper cable required on the vehicle.

Give two reasons why relays are used in motor vehicle circuits:

1 _____

2 _____

On a four-pin relay they are numbered 30, 85, 86 and 87. Each of these pins has a particular connection and understanding these connections and how the relay is energised will assist in diagnosing relay faults.

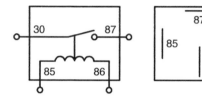

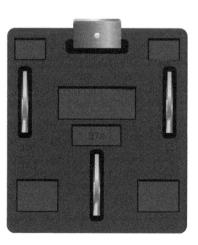

Research the operation of a four-pin relay and identify the pin numbers.

_____

_____

_____

Testing a relay

Testing a relay will help to identify if it is operational or if it requires replacement. To test the relay carry out the following procedure:

1 **Connect battery positive to pin 85.**

2 **Connect a suitable earth to pin 86.**

3 **Once these pins have been connected there should be an audible click – this means the coil has energised and the switch has closed.**

4 **Connect battery positive to pin 30.**

5 **Connect a digital voltmeter (set to 20V DC) to pin 87 and earth and it should read 12V. The digital voltmeter can be replaced by putting a 12V test lamp between pin 87 and earth; if the relay is operating correctly the lamp should illuminate.**

Using the procedure above test a relay using both a digital voltmeter and a test light.

Delay relay

On modern vehicles some electrical circuits encompass a delay relay; a typical use for this would be for the interior lighting circuit. This allows the interior light to stay on for a period of time (set by the delay relay) before self extinguishing. Another type of relay used on modern vehicles is a timer

relay. An example of a timer relay would be on a heated screen. This allows the driver to press the heated screen button and then the heated element switches off automatically after a set period of time which is set by the timer relay.

The diagram below shows a typical wiring diagram of a heated rear screen and heated mirror circuit. Briefly explain the operation of this circuit:

_____

_____

_____

_____

_____

SENSORS

There are two main types of sensors used on vehicles: passive and active sensors. A passive sensor is one that receives signals only and varies its internal resistance, such as a temperature sensor, whereas an active sensor is one that generates an output, such as a crank angle sensor. The output from the sensor can be used to operate actuators to perform a specific task on the vehicle; understanding what is monitored by sensors and what could be operated by actuators will help with fault diagnosis. Using the following list, complete the table to identify where sensors would be used or where actuators are used:

throttle position
idle speed control
dwell angle
engine knocking
variable valve
timing

manifold absolute
pressure
exhaust gas
recirculation
coolant
temperature

mass air flow
turbo boost
crankshaft position
fuel injection
exhaust gas
content

throttle body
camshaft position
ignition timing

Sensor	Actuator
engine knocking	ignition timing

Regardless of the type, sensors indicate electronically a set of values which, if outside pre-set limits, must be corrected. This is done in normal running conditions by the electronic control units (ECUs) to provide efficient operation of the engine, transmission, brakes, suspension etc.; or it is corrected by the technician if a warning indicator operates to signal a fault.

Thermistor

A thermistor is a type of variable resistor, the resistance will change dependent on the temperature. These can be either positive temperature coefficient (PTC) or negative temperature coefficient (NTC). A NTC thermistor's resistance will reduce as the temperature rises which in turn allows more current to flow through. NTC thermistors are commonly used for sensing engine temperature. A PTC thermistor will have a sudden increase in resistance at a preset 'critical' temperature and acts like a switch; these can be used for circuit protection, current limiting devices and over temperature protection.

Temperature (°C)	Typical resistance (Ω)	Actual resistance readings (Ω)
0	4800–6600	
10	4000	
20	2200–2800	
30	1300	
40	1000–1200	
50	1000	
60	800	
80	270–380	
110	0	

Research and describe the action of the following engine management sensors shown:

Knock sensor

Crankshaft position sensor

_____

_____

_____

_____

_____

TRANSISTORS

A **transistor** acts as a 'solid state' switch and can open and close several times per second. There are two types of transistor, NPN and PNP. This describes the construction of the transistor; for example NPN is negative–positive–negative material sandwich and PNP is positive–negative–positive material sandwich. Each of these three sections has a lead connected to it.

This allows any of the three sections to be connected to the circuit. What are the names given to the three legs?

_____

Transistor operation

Transistors are used as a switch or an amplifier. A NPN transistor requires a small voltage of approximately 0.7V to be supplied to the base terminal to allow it to fully switch on. Once switched on the collector and emitter legs can be joined and a larger current can flow through them. It is sometimes easy to think of a transistor as a relay, however, the transistor requires less voltage to operate it. A PNP transistor works in much the same way as a NPN transistor except the voltage required at the base to 'switch on' the transistor is now 0.7V less than the voltage at the emitter, i.e. if the emitter voltage was 1.0V then the voltage required at the base would be 0.3V.

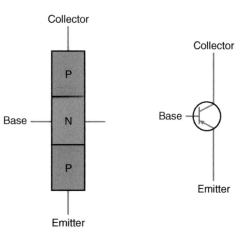

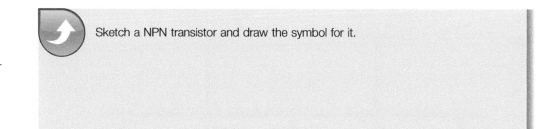

Sketch a NPN transistor and draw the symbol for it.

Transistor acting as a switch

Transistors can be used as a switch to turn items on and off such as LED lights and cooling fans. Describe the operation of the transistor to create current flow and illuminate the light in the circuit shown:

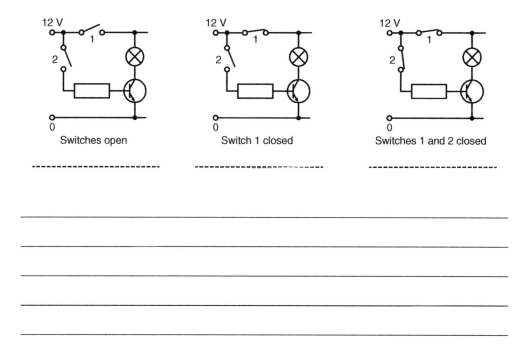

| Switches open | Switch 1 closed | Switches 1 and 2 closed |

---------------------------   -------------------------------   -------------------------

_____

_____

_____

_____

_____

Research what a 'Darlington pair' is and how it works. Sketch a diagram in the space provided to show how a 'Darlington pair' are arranged

_____

_____

_____

_____

Vehicle wiring: conventional multi-wire looms

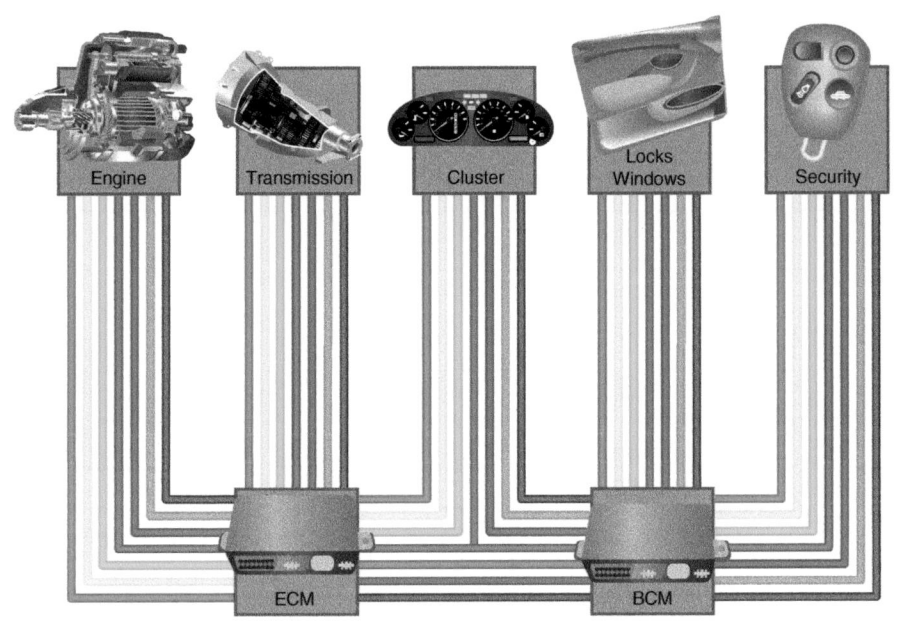

MULTIPLEX

Modern vehicle electrical systems are getting more complex. The number of separate wires on modern high end vehicles can be over 1000. This can cause problems not just in size and weight but due to the increased number of connections and wires it can increase the possibility of faults occurring. To try to overcome this issue multiplex systems were incorporated into motor vehicles; these are sometimes also referred to as data bus systems.

Multiplex systems can use different types of wiring including single wire, twin wire, twisted pairs or even fibre optics. The choice would depend on elements such as the signal speed required, how much signal interference there was and most importantly the financial implications. What would be the most and least expensive types of wires to use for a multiplex system?

_____

_____

Multiplex systems rely on the data bus and power supply cables visiting all areas of the vehicle electrical system. The pictures opposite show the difference in cabling required on a conventional wiring system and a simplified multiplex system.

Vehicle wiring: CAN bus network

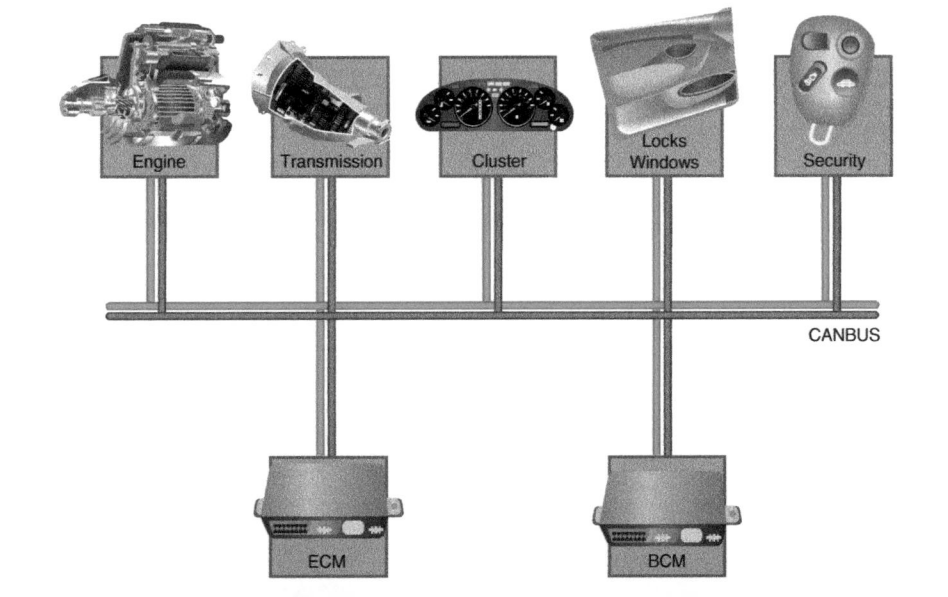

Research and describe what happens in a multiplex system when the driver operates the side light switch.

_____

_____

_____

A data bus carries information from sensors and allows the various ECUs to operate and talk to each other.

The multiplex system is now predominantly made up of two or three controller area network (CAN) buses. These each operate at different speeds and on a two bus system are commonly known as High Speed CAN (HS CAN) and Low Speed CAN (LS CAN). The designations are for transmission protocols and transmission speeds. Systems operating on the HS CAN will have precedence over the systems operating on The LS CAN, for example a signal from the ABS system would take precedence over a signal from the infotainment system. Data transmitted between 100kbits and 1Mbits per second is high speed CAN (HS CAN) and data transmitted between 10kbits and 100kbits per second is low speed CAN (LS CAN).

List four common systems that operate on a low speed CAN bus:

1 _____

2 _____

3 _____

4 _____

List four common systems that would use a high speed CAN bus:

1 _____

2 _____

3 _____

4 _____

When a digital signal is sent on the CAN bus it is made up of several elements, the following shows how a typical message could be sent on the CAN bus:

SOF	MSG ID	RTR	Control	DATA	CRC	ACK	EOF

Identify what each element of the above message means:

SOF _____

MSG ID _____

RTR _____

Control _____

DATA _____

CRC _____

Ack _____

EOF _____

As well as the way data is transferred there are different types of bus operating. The following are examples of these:

Serial bus – Information is transferred by a serial stream of bits, one bit after another.
Multi master bus – peer to peer network, no server, and all control units have the same right to pass data onto the bus.
Message orientated – Messages are broadcast so that all active ECUs receive them and then decide if they are relevant.
Priority controlled – High priority transmissions take precedence over low priority transmissions.

The illustration on page 262 shows a typical vehicle with different bus systems fitted. Give two further examples of systems that would be controlled by each bus system on the following page.

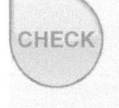

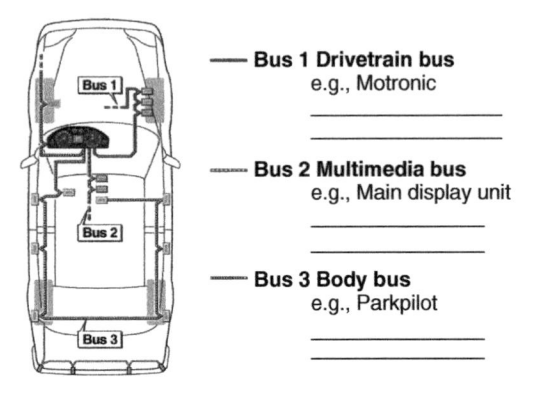

—— **Bus 1 Drivetrain bus**
  e.g., Motronic

_____

_____

------ **Bus 2 Multimedia bus**
  e.g., Main display unit

_____

_____

—— **Bus 3 Body bus**
  e.g., Parkpilot

_____

_____

CANBUS is the circuit that facilitates the transmission of electronic data between modules. The CANBUS system can be made up of one wire or two wires. In a two wire system they are usually twisted and the voltages between the two wires are different, usually either 2 volts or 0 volts. Using the twisted wires with the voltage difference helps reduce the effect of electromagnetic interference including voltage spikes.

In order to further prevent any damage to the CANBUS system from electromagnetic interference, a 120Ω resistor is fitted across the two twisted wires at each end; these are commonly known as termination resistors.

How does a CANBUS system make it easier to diagnose electronic faults on a vehicle?

_____

_____

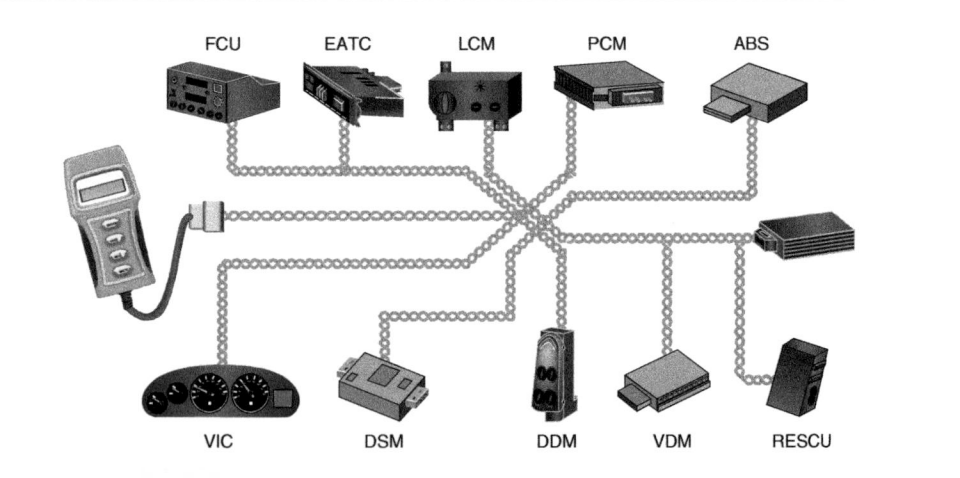

FCU EATC LCM PCM ABS

VIC DSM DDM VDM RESCU

CHECK CAN bus is one of five protocols used in OBD II and EOBD diagnostics. EOBD stands for European On Board Diagnostics

Research the advantages and any disadvantages of using multiplex systems over conventional electrical wiring systems.

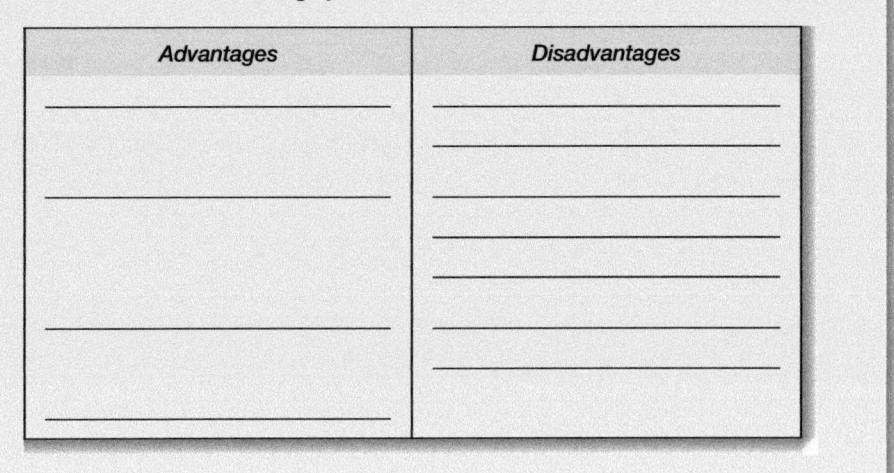

Advantages	Disadvantages
_____	_____
_____	_____
_____	_____
_____	_____
_____	_____
_____	_____
_____	_____
_____	_____

WWW http://canbuskit.com

LIGHTING CIRCUITS AND REGULATIONS

The main purpose of the vehicle lighting system is for the vehicle to be seen and for the driver to be able to see where he is going. The lighting system shows the direction the vehicle is driving and the size of the vehicle. There are some legal requirements regarding the colour of light being emitted. State the legal requirement for the colour of light permitted from the lights on page 263.

Lamps facing rearwards

Lamps facing forwards

Side facing lights and direction indicators

Lighting systems are developing and it is now quite common to see light emitting diodes (LED) being used for some vehicle lighting systems. LEDs used to be used predominantly in brake lights as they illuminated quicker and were brighter than the conventional tungsten bulbs. However, LEDs are used across a range of lighting systems now due to the intensity of the light and they have a greater life expectancy than conventional bulbs; they also operate on lower voltages which in turn reduces the power consumption of lighting circuits.

Headlights

Headlights are mounted at the front of the vehicle and are used to light the road ahead. The headlight system will consist of a dip beam circuit for normal driving and a main beam circuit for when there is poor visibility, however, due to the higher intensity of the main beam they should not be used when traffic is coming towards you as they could blur the vision of the oncoming driver.

Headlights come in a variety of different arrangements. Composite headlight units are still found on many vehicles today. This system has a Halogen headlight with a replaceable bulb. When these headlights were first used they allowed the manufacturers to change the size and shape of the headlight unit which improved the aerodynamics of the vehicle and the driver's visibility.

 When changing a halogen bulb; be careful not to touch the glass envelope as the oil from your skin can greatly reduce the life of the bulb.

To improve driver visibility at night even further, vehicle manufacturers moved to using poly ellipsoid headlights (sometimes known as projector headlights). These headlights are more compact than conventional headlights and use an elliptical shaped reflector instead of the conventional parabolic design. This design allows all the light to be reflected to a single focal point in front of the bulb giving a much more intense and focused light. Poly ellipsoid headlights can use xenon headlights or halogen and in more recent designs LED lights have been used.

Using the poly ellipsoid headlights also allows only one bulb to be used for both dipped and main beam. To achieve this, a shutter is fitted in the headlamp assembly that deflects the light when on dipped beam and the shutter is then retracted when main beam is required giving a full intense forward facing light. What is a disadvantage of using poly elipsoid headlights?

_____

Identify a number of vehicles in the workshop and try to find a vehicle with composite headlights and one with poly ellipsoid headlights. Switch on the headlights and note the difference in the light.

On more modern vehicles there is a newer type of bulb known as HID or xenon. What does HID stand for?

_____

These bulbs are electronically controlled and are identified by the blue–white colour of the light they emit. Instead of using a filament, an electric arc is produced between two electrodes that excite a gas (usually xenon) inside the headlamp. This in turn vaporises metallic salts that sustain the arc and emit light. The presence of the inert gas amplifies the light given off by the arc.

More than 15000 volts are used to jump the gap between the electrodes. However, once the gap has been bridged the bulb only requires about 80 volts to keep the current flowing. In order to produce this starter voltage a voltage booster and controller (known as a ballast unit) is required in the system. Once the headlights are switched on it can take up to 15 seconds for the light to reach its maximum intensity. One of the main advantages of xenon headlights is that the light emitted is brighter and more consistent than conventional halogen headlamps. However, one of the disadvantages of xenon headlights is that they are electronically operated and therefore may require coding into the vehicle ECU when replacing.

 When replacing xenon headlights always make sure they have had time to cool down and the electrical supply to the headlight is disconnected.

Headlamp alignment

All vehicle headlamps in the UK must comply with the Department for Transport Road Vehicles Lighting Regulations, which state the position, or angle, of the dipped beams. The specialist equipment used to check the alignment of headlamps measures the angle of the dipped beams and the beams' position relative to one another when on both main and dipped beams.

The diagram below shows the correct pattern and tolerances for a dipped beam headlight.

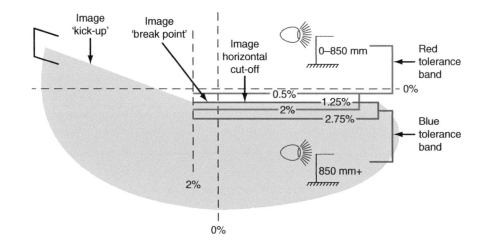

Draw the correct headlight beam pattern for main beam on the aiming grid below; include the tolerances as per MOT standards.

1. Typical main driving beam position all types

Vertical screen | Aiming line

Horizontal screen

Aiming line

http://www.ukmot.com/mot%20manual/?MOT=Headlamp%20Aim&MOT_Number=1.8&MOT_Section=lights

TIP Asymmetric headlights are the standard for European headlights and symmetric headlights are the standard for US headlights.

Check the headlamp alignment of a vehicle using available equipment and describe the alignment procedure:

● Vehicle make and model:_____

_____

● Pre checks:_____

_____

● Procedure:_____

_____

● Where would you find the headlamp alignment settings for the chosen vehicle?

_____

_____

Lighting circuits

Headlight wiring diagrams can be very complex, understanding these diagrams and being able to follow the different circuits will make diagnosing faults a lot easier. The diagram on page 267 shows a typical headlight and fog light system. In groups, identify each circuit and explain them to your assessor/tutor.

Using a manufacturer's technical manual find the headlight wiring diagram and explain the circuit to your assessor/tutor.

DIAGNOSTICS: LIGHTING – SYMPTOMS, FAULTS AND CAUSES

State a likely fault and cause for each symptom listed below. Each cause will suggest any corrective action required.

Symptoms	Faults	Probable cause
Intermittent light operation: inoperative lamp	Loss of earth	_____
Intermittent lamp operation/ dim light	_____	_____
Headlamp beam pattern incorrect	_____	_____
Short circuits: Ingress of water	_____	_____
Indicators not working correctly	_____	_____
Low intensity light	_____	_____
Inoperative light	_____	_____

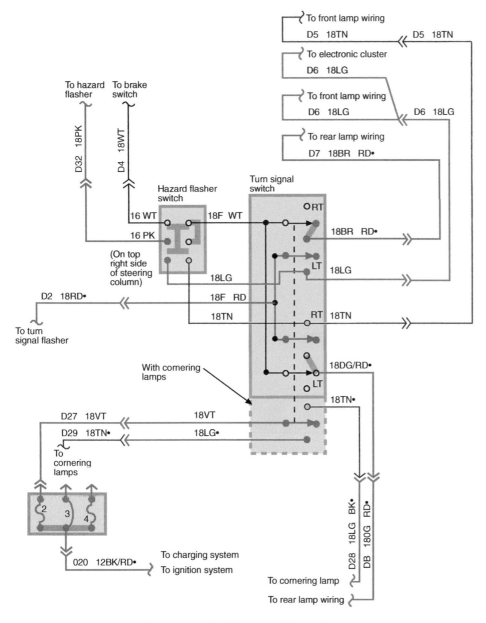

To front lamp wiring
D5 18TN — D5 18TN

To electronic cluster
D6 18LG

To front lamp wiring
D6 18LG — D6 18LG

To rear lamp wiring
D7 18BR RD•

To hazard flasher   To brake switch

D32 18PK   D4 18WT

Hazard flasher switch

Turn signal switch

16 WT   18F WT

16 PK

(On top right side of steering column)

18LG

D2 18RD•   18F RD
18TN

To turn signal flasher

○ RT
18BR RD•

LT
18LG

RT 18TN

With cornering lamps

18DG/RD•

○ LT

18TN•

D27 18VT   18VT
D29 18TN•   18LG•

To cornering lamps

2   3   4

020 12BK/RD•   To charging system
To ignition system

D28 18LG BK•
DB 180G RD•

To cornering lamp
To rear lamp wiring

Turn signal circuit for a two-bulb system.

INFOTAINMENT SYSTEM

Entertainment systems in vehicles have improved dramatically over recent years, no longer is the radio cassette player the standard equipment. Infotainment systems can now include CD players, connections for MP3 players, Bluetooth connections for hands free phone use, satellite navigation systems and even DVD players

The picture below shows a standard entertainment system that can be found in a lot of vehicles. This unit allows six CDs to be loaded into the front of the CD player and can choose music from different CDs randomly.

Car stereo

The removal of infotainment systems is quite complex in some vehicles and requires specialist tools. This is a design feature to try to reduce or even prevent infotainment system theft. The image on page 268 shows a set of specialist tools required for removing infotainment systems.

_____
_____
_____

Optional gasket —————————————————————— Gasket
Frame ——————————
Front plate ——————— Surround
Magnet ———————
Back plate ————— Cone
Center pole ————— Spider
Vent ————— Screen
Voice coil & former ————— Dust cap
Connection terminal —————

Some car stereos require a code to operate them, ensure this code is known prior to removing the stereo where applicable.

Speakers

Modern in-car entertainment systems use at least four speakers and some vehicles will have in excess of ten speakers. When fitting speakers it is important to ensure the output of the speaker is matched to the entertainment systems amplifier or it may cause the sound to be distorted.

The sound quality of an entertainment system is largely dictated by the quality of the speakers and their match to the system.

Explain how speakers operate.

_____
_____

Low frequency sounds require a speaker with a large diaphragm and a big mass, while high frequency sounds require a speaker with a small diaphragm and a low mass. Because of this there are two main types of speakers known as tweeters or woofers. To cover the full range of frequencies required a combination of both types are commonly used. Woofers operate in the frequency range of 40Hz to 2000Hz (although they can go up to 5000Hz in some cases) and tweeters operate in the frequency range of 2000Hz to 20000Hz.

Tweeters are for: _____

Woofers are for: _____

Tweeters   Woofers

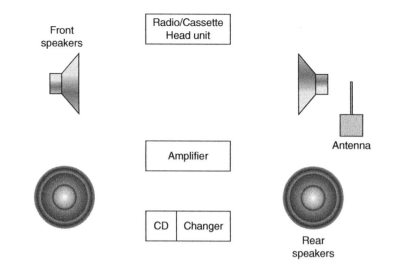

The reception and reproduction of sound by an entertainment system relies not only on the quality of the equipment but also on the quality of the installation. Most reception and sound deficiencies are due to problems with the installation. It is important to note that when carrying out any repairs to the wiring on infotainment systems the connections should always be soldered and not joined by makeshift connectors. This is to ensure the signals remain as clear as possible and to prevent any interference and degradation of sound.

 Take some time to practise soldering wires together and then check continuity and resistance using a multimeter.

Complete the drawing above right to show the installation wiring for a radio and list the important points to be aware of when installing a radio:

When fitting a DVD player into the front passenger compartment of a vehicle it should not operate unless the handbrake is applied, for safety reasons. Therefore it should be linked to the handbrake warning light system.

Due to the complexity of some more modern infotainment systems there is a need for it to run on its own multimedia bus system to prevent interference and to allow data transmission to be as quick as possible. A common multimedia bus system is MOST; this is a multimedia network that provides a cost effective interface for all multimedia components in the vehicle. What does MOST stand for?

MOST – _____

The diagram below shows the components that can be linked through the MOST network:

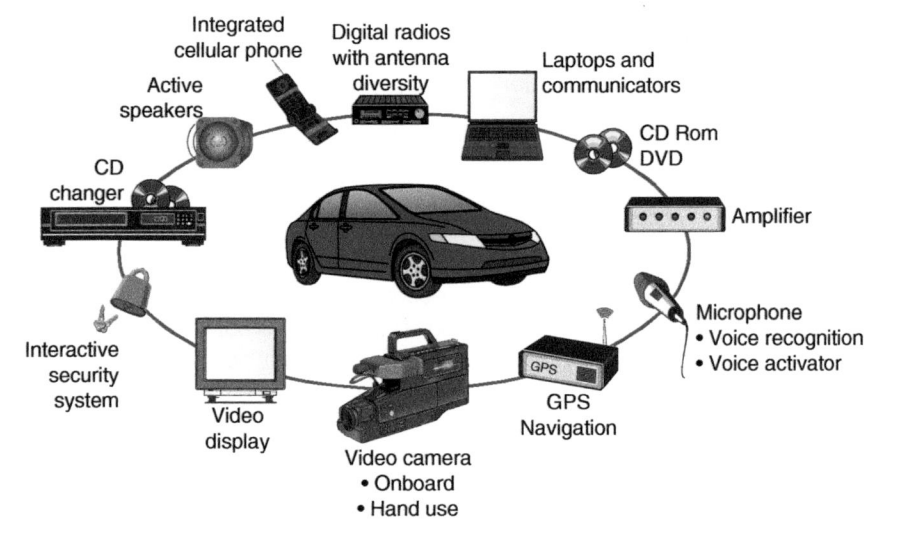

Using a multimedia network allows entertainment systems to be added with the minimum disruption to those already fitted; however, when changing items they will require coding into the network to ensure they operate correctly. Operating the multimedia system on its own network can also aid in faster fault diagnosis on infotainment systems.

Mobile communications

Prior to the onset of Bluetooth, phones could be hard wired into the entertainment system and in some cases these are still available. However, more often in modern vehicles mobile phones can be connected to the in-car entertainment system via Bluetooth. Once the phone is connected via Bluetooth it can use the radio speakers as the phone speaker and there is a small microphone fitted to give hands free speech. When the phone is in use it overrides the stereo and therefore background noise is eliminated.

Research how Bluetooth operates.

_____

_____

_____

Due to the rapid development in mobile communications equipment there is sometimes difficulty with connecting the most modern phones with some vehicles. This is because a vehicle takes some time to develop and it may still have an older version of the Bluetooth technology. How can this difficulty connecting a mobile phone to a vehicle be overcome?

_____

_____

Navigation systems

Vehicle navigation systems are becoming more popular and in some cases are fitted as standard to modern vehicles. The system helps drivers to navigate between two different points by using a network of Global Positioning System (GPS) satellites which are continually orbiting the Earth. This network of satellites can accurately identify locations to about 10–15m. Each satellite sends out data which is received by the vehicle and the system then uses a mathematical algorithm to locate its position and plot it on a grid stored in the GPS.

The vehicle position can be plotted on this grid the grid stored in the GPS and therefore the system knows where the vehicle is located. The vehicle navigation system then plots the location onto the digital maps either stored in its memory or on disk and this is shown as a location on the map screen.

To further improve the accuracy of locating the vehicle, the system also uses signals from ground based mobile phone towers and radio masts. If the navigation system is fitted to the vehicle systems it will also use data from the vehicle, such as speed and distance travelled, to make the tracking of the vehicle location more accurate, especially in built up areas where the GPS Satellite signals may be weaker due to interference from buildings, bridges and tunnels. A typical in-car navigation system can be seen on page 271.

Most modern GPS systems are accurate to within ± 10m and although earlier models did not perform too well in forested areas or near tall buildings, more modern systems do not have the same problems. GPS satellites transmit signals using radio waves which are received by the GPS system in the vehicle. To determine the position of a vehicle the satellite records how long it takes for the signal to reach the GPS in the vehicle; this must be very accurate timing so the satellites use atomic clocks to achieve this. In order to give a two-dimension location of a vehicle a minimum of three satellites are required.

Each satellite sends a signal and the GPS receives them at slightly different times due to the distance of the satellites from the vehicle. These satellites each give an area the vehicle could be in, the point at which all three areas intersect is where the vehicle is located. This is known as triangulation. If a fourth satellite is added this can now give a three-dimensional location which includes altitude. The more satellite signals that are used the more accurate the location of the vehicle.

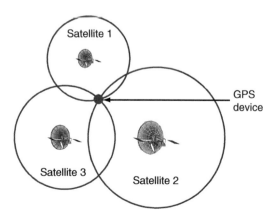

CHECK   A minimum of three satellites are required to locate a vehicle's position – this is known as triangulation; however, a fourth satellite can give a far more accurate location when the vehicle is moving.

Explain how a navigation system can be used? What are its key functions?

_____

_____

_____

_____

Research how to set up a new satellite navigation system and record any consideration that may need to taken into account.

_____

_____

_____

_____

DAB radio

Digital audio broadcasting (DAB) is digital radio. DAB uses digital radio signals instead of analogue signals which give clearer sound quality, more services, less interference and less signal fade which means no need to retune during travel. DAB can also display information on the radio head using RDS technology. The information can be the radio station, what song is playing and the singer. What does RDS stand for?

RDS _____

Most RDS radios also incorporate a function known as EON which allows traffic announcements to interrupt the normal broadcast station to give 'live' updates on traffic from local stations. This feature will also interrupt a CD or tape that is being played to give the announcement. Once the announcement has finished the stereo will automatically return to either the original station being listened to or the CD or tape. What does EON stand for?

EON: _____

Multiple choice questions

Choose the correct answer from a), b) or c) and place a tick [✓] after your answer.

1 **Which of the following is NOT a true statement about LEDs?**

   a) LEDs have a shorter operating life than conventional bulbs [ ]

   b) LEDs can produce a brighter light than conventional bulbs [ ]

   c) LEDs can operate quicker than conventional bulbs [ ]

2 **The stoplight switch is usually operated by the**

   a) Transmission [ ]

   b) Brake pedal arm [ ]

   c) Instrument panel [ ]

3 **How many satellites are required to locate a vehicles position?**

   a) one [ ]

   b) Two [ ]

   c) Three [ ]

4 **The three legs of a transistor are called**

   a) Emitter, box, collector [ ]

   b) Emitter, base, collector [ ]

   c) Emitter, base, cover [ ]

5 **A Hall effect sensor would produce what type of waveform?**

   a) Sine wave [ ]

   b) Half sine wave [ ]

   c) Square wave [ ]

SECTION 2

Starting and charging systems

USE THIS SPACE FOR LEARNER NOTES

_____

_____

_____

_____

_____

Learning objectives

After studying this section you should be able to:

● Identify the different types of starter motor.
● Describe the operation of different types of starter motor.
● Determine how to diagnose faults on starter systems.
● Identify charging system components.
● Describe the construction and operation of charging system components.
● Determine how to diagnose faults on charging systems.
● Describe the operation of start/stop systems.
● Identify components of start/stop systems.

Key terms

Pinion A small gear on the end of the starter motor which meshes with the ring gear.
Solenoid A heavy duty electromagnetic switch which can be used in a number of applications, including the starter motor.
Rectifier The component that changes alternating current to direct current.
Diode An electronic component that allows current to flow in one direction only.

www.autotalk.co.uk

www.alternatorparts.com

http://electronics.howstuffworks.com

www.allaboutcircuits.com

TYPES OF STARTER MOTOR

A starter motor has one basic functional requirement, explain this requirement below:

_____

_____

The operation of the starter motor is based on the motor converting electrical energy (battery voltage) into mechanical energy (turning motion of the pinion).

The different sizes and types of engines make it necessary to require different types of starter motor.

Name the different types of starter motor shown below.

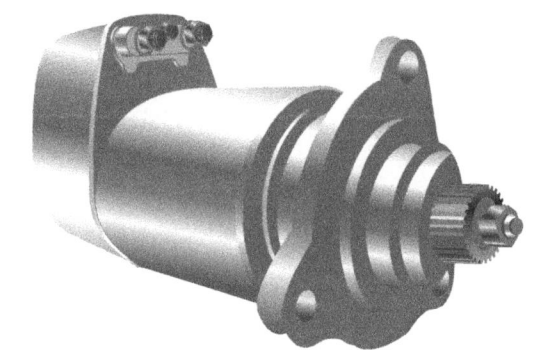

_____

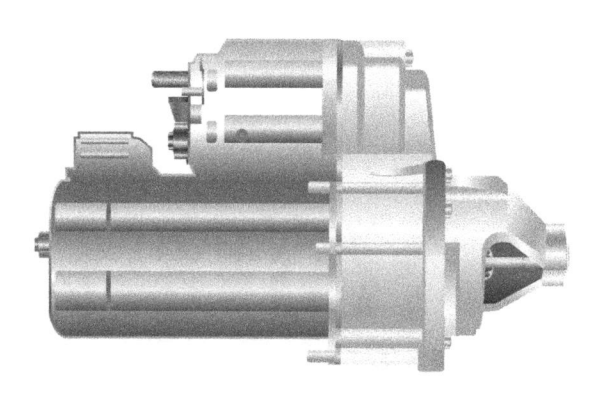

Starter motor wiring

The wiring on modern starter motors is fairly straightforward. However, if it is wired incorrectly it can cause catastrophic failure. There are three main connections to consider: M, S and B as shown in the diagram below. What is connected to each of these terminals?

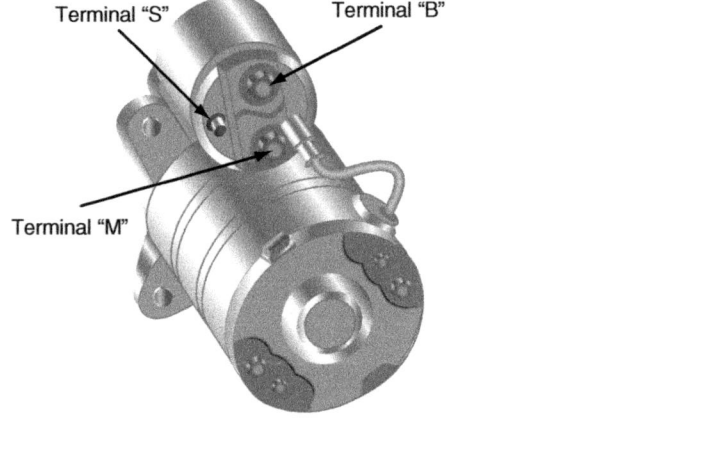

Terminal "S"   Terminal "B"

Terminal "M"

B – _____

S – _____

M – _____

Starter motor operation

The starter motor is designed to turn the engine fast enough until it can run under its own power. To do this a large amount of current is drawn from the battery to the starter motor; this can be as much as 250 amps or more on a large starter motor. In order to allow this amount of current to be drawn, large thick cables are required.

To allow the cables to be connected directly to the starter a control circuit is operated via the ignition switch, starter relay and starter solenoid. All these components use smaller wires to operate them and only once the relay and solenoid are energised does the high current pass down the large cable from the battery to the starter.

The following is a diagrammatic view of a basic starter motor wiring circuit. With the aid of this diagram explain what happens during starting.

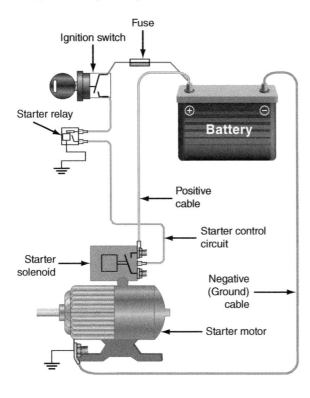

_____

_____

_____

_____

Once the engine has started and the engine speed is increasing, the starter motor could become damaged if it remains connected and spinning. To prevent this damage an over-run clutch is fitted. This clutch locks the pinion to the armature shaft when the armature shaft is rotating and,

in turn, this allows the pinion to turn the ring gear. When the engine starts to turn faster than the armature the over-run clutch releases and allows the pinion to turn freely on the armature shaft.

Starter pinion gear

 Disconnect the battery to prevent short circuits with spanners when working on starter systems.

PERMANENT MAGNET STARTERS

Starter motor technology has changed over the years and technology has progressed. More modern vehicles are now using starter motors with permanent magnets rather than the conventional electromagnets. What is the advantage of this type of motor?

_____

_____

State three advantages of using permanent magnet starters:

1 _____

2 _____

3 _____

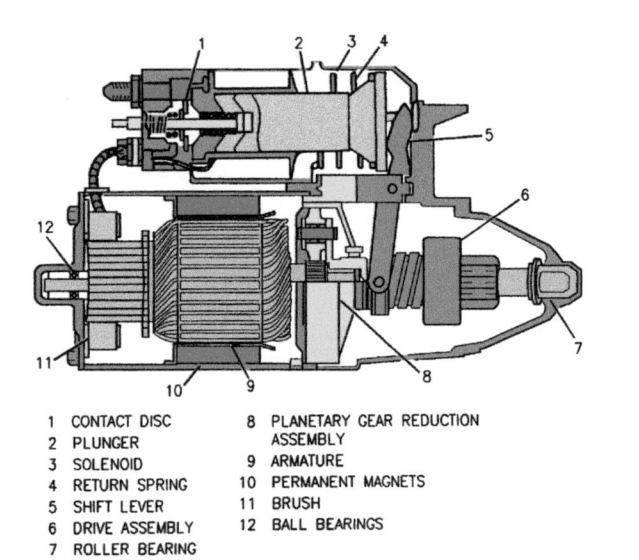

1 CONTACT DISC
2 PLUNGER
3 SOLENOID
4 RETURN SPRING
5 SHIFT LEVER
6 DRIVE ASSEMBLY
7 ROLLER BEARING

8 PLANETARY GEAR REDUCTION
   ASSEMBLY
9 ARMATURE
10 PERMANENT MAGNETS
11 BRUSH
12 BALL BEARINGS

 **TIP** Handle permanent magnet starter motors with care. Avoid dropping them or sharp blows to the casing as the brittle magnet material is easily damaged.

TESTING/DIAGNOSING STARTER MOTOR FAULTS

If the starter motor is considered to be faulty a systematic check should be carried out to determine where the fault lies. Prior to carrying out a systematic check there are a number of preliminary checks that should be completed. List the preliminary checks that should be made:

- _____
- _____
- _____
- _____

Once the preliminary checks are carried out satisfactorily the systematic checks can now be completed.

Cranking voltage test

This test will measure the amount of voltage available to the starter motor during cranking. Connect a remote starter switch between the battery positive terminal and the starter terminal on the starter motor. Connect a voltmeter (set to 20V DC) between a good earth and the starter motor feed. Operate the remote starter and record the battery voltage as the engine turns; this should not be less than 9.6 volts. State the likely problems if the voltage is high or low:

High: _____

Low: _____

Label the diagram below.

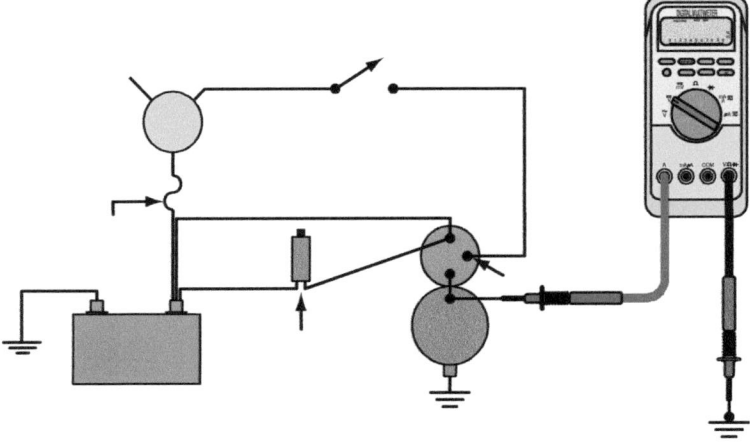

Cranking current test

This test will measure the amount of current the starter motor draws during cranking. With the remote starter switch and voltmeter still connected, also connect an amp clamp around the battery negative lead. Disable the ignition system and crank the engine over for a maximum of

15 seconds. Record the voltmeter reading and amp clamp reading. State the likely problems if the current draw is high or low:

High: _____

_____

_____

_____

Low: _____

_____

_____

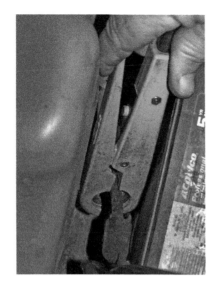

Jump cable

Starter relay by-pass test

This test will confirm if the relay is serviceable or not. Disable the ignition system and then connect a suitable jump cable between the battery positive terminal and the starter terminal on the starter relay. Once the connection is made the starter should turn. If the starter motor turns correctly then the starter relay is faulty.

Earth circuit resistance test

This test will check the earth return from the starter motor. Connect a remote starter switch as detailed previously. Connect the voltmeter to the battery negative and the starter motor casing. Operate the remote starter switch and record the voltage drop. This should be no more than 0.2 V for a 12 volt vehicle.

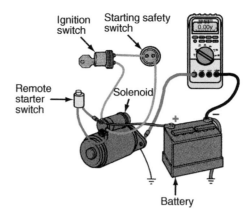

 Carry out the previously described tests on a starter system and record the findings in the table below. Compare the findings with manufacturer's data and give a conclulsion on the serviceability of the starter system.

Test	Actual readings	Manufacturer's recommended data
Cranking voltage test		
Cranking current test		
Starter relay by-pass test		
Earth circuit resistance test		

 Before carrying out any of the above tests ensure the vehicle is in neutral.

 A quick check to confirm if the starter circuit is faulty or the battery is flat, is to turn the vehicle headlights on, operate the starter and if the lights go dim then the fault lies with the battery. If the lights do not dim the fault is in the starter circuit.

Bench testing a starter motor

If a starter motor is removed from a vehicle it can be tested on a bench in a controlled manner. This will test the starter motor with no load applied.

To conduct the test the starter motor will need to be secured in a vice. Once secured take a known good battery and connect a jump lead from the negative battery terminal to the starter motor casing. Connect a jump lead from the battery positive terminal to the battery positive terminal on the starter motor solenoid. Now connect a remote starter switch between the battery positive terminal on the starter motor solenoid and the switch terminal on the starter motor solenoid and operate the switch. The starter motor should now turn freely.

State what conclusions can be made from carrying out a bench test of a starter motor:

- _____
- _____
  _____

- _____
  _____

 Carry out a bench test of a starter motor and report on its condition.

CHARGING SYSTEMS

The type of charging system found on all modern vehicles consists of an alternator and storage battery. State the purpose and functional requirements of (a) the charging system and (b) the battery:

a _____
  _____
  _____

b _____
  _____
  _____

The alternator is rotated by a drive belt driven from the engine. From this rotation and through the internal components the alternator produces an electrical output. The rotor spins inside a stator to produce an alternating current (AC) voltage; however, this AC voltage is of no use to the vehicle electrical systems or for charging the battery. The alternator has a **rectifier** pack made up of several **diodes** which changes the AC voltage to a direct current (DC) voltage which can now be used on the vehicle electrical systems. To ensure the voltage produce is not too high for the vehicle systems, a voltage regulator is fitted. This will control the alternator output to between 13.6 volts and 14.7 volts DC.

Correctly label the exploded view of the alternator using the terms listed below:

voltage regulator and brushes	bearings	cooling fans	stator
rectifier assembly	rotor	end cover	
	slip rings	drive pulley	

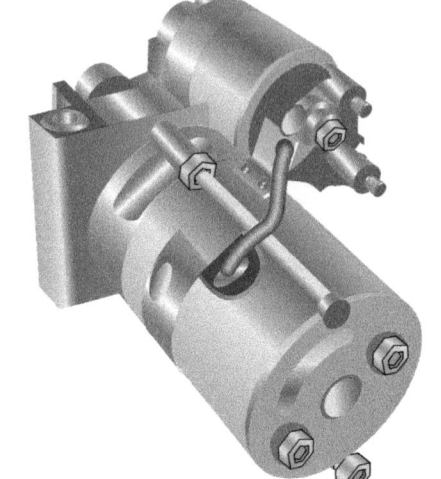

When the engine is running keep fingers and loose clothing away from moving components, e.g. alternator drive belts.

Alternator wiring

The wiring on modern alternators can be either in an electrical block connector or using electrical post terminals. The connections may look different, however, they are all connected to the same parts of the charging circuit. The diagrams below show a typical three connection alternator and a typical four connection alternator. Knowing these terminals can help with performance testing and fault diagnosis.

Three connector alternator

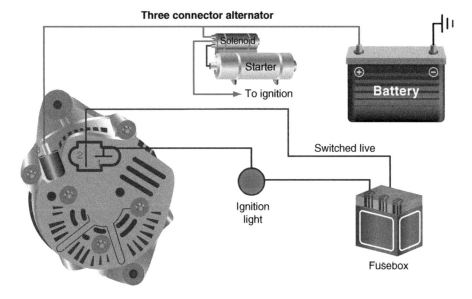

Four connector alternator (b)

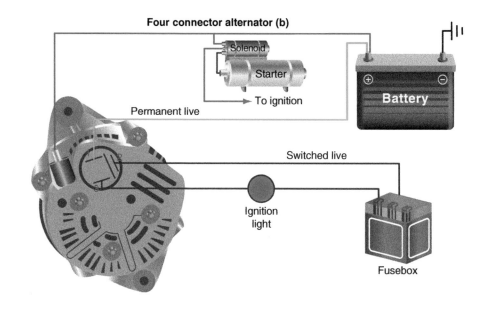

Select three different alternators and identify which terminals correspond to those shown in the diagrams above.

TESTING/DIAGNOSING ALTERNATOR FAULTS

If the alternator is considered to be faulty a performance check should be carried out to determine where the fault lies. Prior to carrying out this check there are number of preliminary checks that should be completed. List the preliminary checks that should be made:

- _____

- _____

- _____

- _____

- ● _____

- ● _____

After completing the preliminary checks, carry out the basic alternator tests shown below.

 **TIP** Before using electrical test equipment always check it for condition and calibration to ensure it will give accurate readings.

Noise checks

Listening to the alternator when the engine is running can determine certain faults. The most common noise heard is a screeching sound on start up that gets louder and more pronounced as the engine speed increases. What is the most likely cause of this noise?

_____

The other common noise on alternator operation is a whirring/whining noise which again gets more pronounced as the engine speed increases. What is the most likely cause of this noise?

_____

Voltage output checks

It is essential that the alternator voltage output can match the demands of the vehicle electrical system without causing damage to components. To test the voltage output, connect a voltmeter (set to 20V DC) across the battery terminals. Check and record the battery voltage with the engine turned off. Start the engine with all electrical components turned off and run at the manufacturer's recommended rpm (usually 1500rpm) and record the voltage displayed on the voltmeter; this should be between 13.2 and 14.7 volts DC. If the voltage is below 13 volts this indicates a charging fault. If the voltage is above 16 volts this indicates an overcharging fault. State a fault that could cause low voltage output and high voltage output:

Low: _____

High: _____

If the results of the test are within limits and the alternator is still suspected of being faulty, increase the engine speed to around 2000rpm (check manufacturer's data) and put on the headlights and other high current electrical components such as the heated rear screen, and record the voltage on the voltmeter. Compare this with the manufacturer's specifications. The voltage should be around 13 volts.

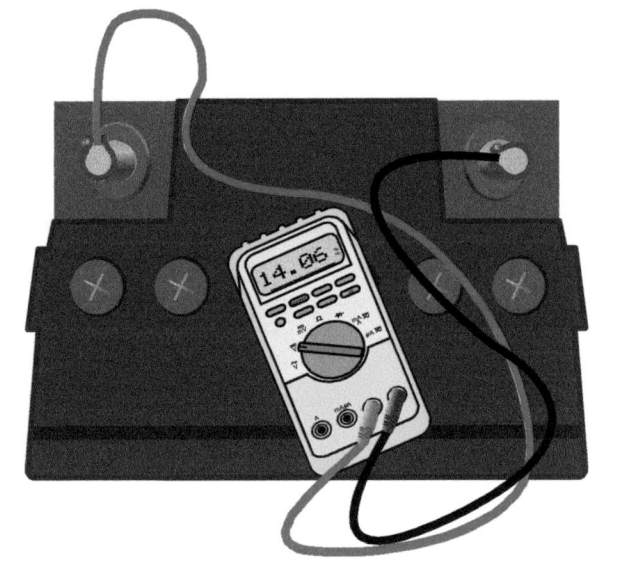

Current output check

The current output of the alternator is very important as it is the current that electrical components require to make them operate. To carry out this test ensure all electrical items are turned off and secure an amp clamp around the battery negative lead. Start the engine and record the current draw, compare this with the manufacturer's data. Turn on electrical items that use a lot of current such as headlights, airconditioning, heated rear screen and wipers and record the current reading. The more items that are turned on the higher the current draw should be. Compare these figures with the manufacturer's data.

What reading will indicate that the alternator is faulty?

_____

_____

AC leakage test

AC voltage cannot be used by the vehicle electrical system to charge the battery or to run electrical items, therefore it is important that no AC voltage is leaking past the rectifier diodes. To carry out this test connect an ammeter in series with the battery positive post of the alternator with the engine switched off. Set the ammeter to milliamps and the reading should be less than 0.5 milliamps. If the reading is higher then the rectifier pack will require changing.

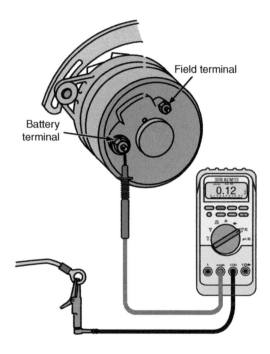

Field terminal

Battery terminal

Carry out the above tests on an alternator and record the findings in the table opposite. Compare the findings with manufacturer's data and give a conclusion on the serviceability of the charging system.

Test	Actual readings	Manufacturer's recommended data
Voltage output checks (no load)		
Voltage output checks (load)		
Current output check (no load)		
Current output check (load)		
AC leakage test		

The B+ terminal is always live on the alternator even when the ignition is switched off. Take care not to short this terminal to earth. For this reason it is important to always disconnect the battery before removing the alternator.

LIQUID COOLED ALTERNATOR

Recent developments in alternator design have produced liquid cooled alternators. By using a liquid cooling system, it is easier to keep the diodes in the rectifier at the optimum temperature for operation. Using the cooling system to keep the alternator cool allows the cooling fans to be removed and therefore reduces alternator noise. It is also expected that due to the more efficient controlling of the alternator temperature the components will last longer.

SMART CHARGING SYSTEMS

A further advancement in charging technology is the 'smart' charging system. With a conventional charging system the alternator output is controlled by the voltage regulator; however, with the 'smart' charging system the alternator output is controlled by an electronic control module. This allows the output to be controlled by the battery temperature, engine temperature and electrical demand. It is temperature dependent because a cold battery responds better to a higher voltage and a hot battery responds better to a lower voltage. Controlling the output to the demands of the battery and vehicle makes the system more efficient.

STOP/START TECHNOLOGY

There is a drive in current technology to increase fuel economy, reduce emissions and make vehicles more reliable. Many people who drive vehicles in heavily congested areas are continually stopping, starting and standing still. All the time the engine is operating it is using fuel and producing emissions. One way to overcome this problem is by using a technology that can switch the engine off and on depending on how the vehicle is operating. The technology used is known as stop/start technology and is used in vehicles which are known as mild hybrids. This technology is readily available in a wide range of vehicles due to engineers applying advanced electronics to starters and generators.

What effect has this system had on the requirements for the vehicle battery and starter?

_____

_____

_____

_____

The battery in a stop/start vehicle will need to be more robust as it is required to cope with the additional amount of engine starts and will also need to be able to provide sufficient power for ancillary electrical equipment when the engine is stopped.

What is used to stop the battery from losing its charge?

_____

_____

_____

_____

Stop/start systems also require information from other vehicle sensors such as wheel speed sensors, gear lever position sensors and clutch/brake pedal position sensors to ensure the system only operates under certain conditions.

Starter-generator

Some manufacturers use an integrated starter-generator (also known as a starter-alternator) in their stop/start systems. This component replaces both the starter motor and the alternator on a conventional engine.

How does the starter-generator work?

_____

_____

_____

The picture on page 283 shows a switched reluctance starter-generator. Where is a starter-generator located?

1 _____

_____

2 _____

_____

Stator

Rotor

If the stop/start system fails to operate the most common cause is a low charge in the battery. Other reasons include DPF regeneration and auxiliary electrical demands such as airconditioning. When working on this system it is essential that manufacturer's diagnostic techniques are followed as the systems differ quite considerably.

Always follow manufacturer's recommended instructions when working on or near high voltage motor/generators and their related systems.

Operation of stop/start system

A common mode of operation for a stop/start system on a vehicle with manual transmission can be explained as follows:

When the vehicle is at very low speeds or at a standstill the driver depresses the clutch pedal, selects neutral and lifts his foot off the clutch pedal. This will stop the engine; however, the ignition will remain on. When the driver is ready to move again he depresses the clutch pedal and the engine will restart.

Explain a common mode of operation for a vehicle with an automatic transmission:

_____

_____

Multiple choice questions

Choose the correct answer from a), b) or c) and place a tick [✓] after your answer.

1 **Vehicles with stop/start systems on them are also known as:**

 a) Full hybrid [ ]

 b) Half hybrid [ ]

 c) Mild hybrid [ ]

2 **What part of an alternator is the rotating magnetic field?**

 a) Stator [ ]

 b) Brushes [ ]

 c) Rotor [ ]

3 **The alternating current produced by an alternator is converted to direct current through the use of:**

 a) Transistors [ ]

 b) Diodes [ ]

 c) Capacitors [ ]

4 **The device that prevents the engine from turning the armature in a starter motor is called the**

 a) Flywheel [ ]

 b) Pinion gear [ ]

 c) Over-run clutch [ ]

5 **Which one of the following starter motors is likely to be found on a modern light vehicle?**

 a) Inertia [ ]

 b) Coaxial [ ]

 c) Pre-engaged [ ]

6 **Which one of the following, controls the maximum alternator output?**

 a) Regulator [ ]

 b) Rectifier [ ]

 c) Stator [ ]

SECTION 3

Vehicle body electrical systems

USE THIS SPACE FOR LEARNER NOTES

_____

_____

_____

_____

_____

Learning objectives

After studying this section you should be able to:

- Describe the construction and operation of electric windows and mirrors.
- Describe the operation of screen heating systems.
- Describe the operation of wiper systems.
- Describe the operation of locking systems.
- Describe the operation of heating, cooling and air conditioning systems.

Key terms

R134a   Refrigerant used in modern mobile air-conditioning systems.
Transponder   An electronic device used to wirelessly transmit and receive electrical signals.
Reversible motor   A motor that can operate in both directions by changing the polarity to the motor.
Wiper linkage   The mechanical device that connects the wiper motor to the wipers.
Potentiometer   A variable resistor used to control voltage.

www.wisegeeks.com

www.airconcars.com

www.howstuffworks.com

www.allaboutcircuits.com

AUXILIARY ELECTRICAL SYSTEMS

Auxiliary electrical systems fitted to vehicles include the following:

1   *Windscreen and headlamp wipers and washers*

2   _____

3   _____

4   _____

5   _____

6   _____

WINDSCREEN WIPER SYSTEMS

Statutory regulations require that a road vehicle should be equipped with one or more windscreen wiper and washer to give the driver a good view of the road ahead in all weather and driving conditions. A relatively powerful motor is required to drive the mechanism and it is desirable that the motor and mechanism are quiet in operation.

The electric motor may be a single, twin or variable speed type. The combination of the motor and the gear reduction in the wiper motor linkage ensure the smooth operation of the wiper system.

In order to diagnose faults on the wiper system it is important to understand the system. For example, if the wipers are operating slowly, a check that can be carried out to identify if the fault lies with the linkage or the motor is to disconnect the linkage from the motor and operate the wiper switch. If the motor is rotating freely and at a suitable speed the fault lies in the linkage system; however, if the motor does not operate correctly then it is either an electrical fault or a faulty motor. To identify electrical faults it is imperative the technician can follow a wiring diagram and understand what readings there should be at different points in the system. The wiring diagram above right shows a basic two-speed wiper motor circuit utilising two 5-pin relays.

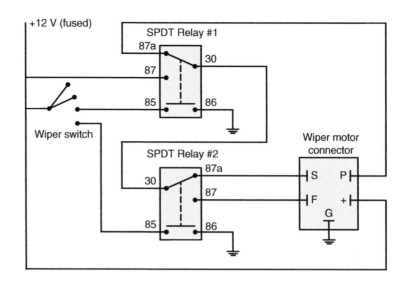

Slow speed operation

When the wiper switch is moved to the slow speed position, 12 volts are supplied to terminal 85 of relay number 1. This now completes the control circuit with terminal 86 and the relay is energised moving the relay from terminal 87a to terminal 87. Twelve volts now pass through terminal 87 to terminal 30 and then onto terminal 30 of relay number 2. As relay number 2 is not energised the 12 volts from terminal 30 now pass through to terminal 87a and then onto the slow speed connector (S) on the wiper motor connection and in conjunction with the battery positive (+) feed and the earth (G) on the wiper motor connection this allows the wiper motor to operate in the slow speed.

Using the diagram explain the operation when the wiper switch is moved to fast speed:

_____

_____

_____

_____

_____

_____

_____

_____

_____

_____

Almost all types of wiper motor have a limit switch device incorporated into the drive to allow the wipers to self park. The switch can operate off a cam or latch arm which supplies voltage to the motor after the wipers have been turned off to allow them to keep running until they reach the park position. The image below shows a wiper motor with the park switch circled:

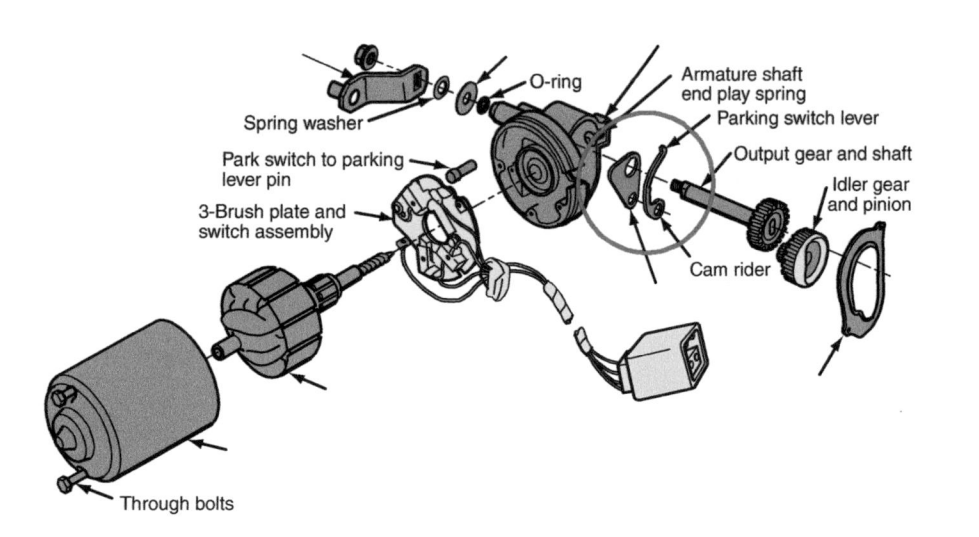

Label the parts of the diagram indicated above.

Intermittent wiper systems

Most vehicles will have a wiper system that has a low speed, high speed and intermittent setting. When the low or high speed position is selected the wipers will operate continuously. When the intermittent speed is selected it allows for one sweep of the wipers and then an interval before the next sweep. On older vehicles the delay between wiper sweeps was set at a pre-determined time by the manufacturer. This technology was further developed to have a number of discreet settings

with different time delays between wiper sweeps to accommodate the differing nature of the rain fall and vehicle speed i.e the faster the vehicle was moving the less time was required between each wiper sweep. This technology was improved even further by fitting a driver adjusted intermittent switch. This switch uses a type of variable resistor called a potentiometer.

What system is used to discharge and start each wiper sweep?

_____

Explain what happens when the driver changes the position of the intermittent switch on a sliding scale wiper system. What is the advantage of this system?

_____

_____

_____

A customer complains that the wipers are not working when they operate the wiper switch. Describe the checks to carry out to identify if the wiper motor is faulty.

- _____
- _____

_____
_____
_____
_____

- _____
_____
_____

Rain sensing wipers

Wiper technology has improved vastly over recent years and the newest technology utilises rain sensing wipers. This system means the driver no longer has to worry about which setting to have the wipers on as the vehicle will automatically adjust the wipers to the conditions.

Explain how rain sensing wipers work.

_____
_____

_____
_____
_____
_____
_____
_____

Most rain sensing systems have a safety feature fitted which requires the rain sensing action to be activated each time the vehicle is switched on to prevent the system from operating with a frozen windscreen or when in a car wash; both of which could cause damage to the wiper system. The diagram below shows the basic operation of a rain sensor.

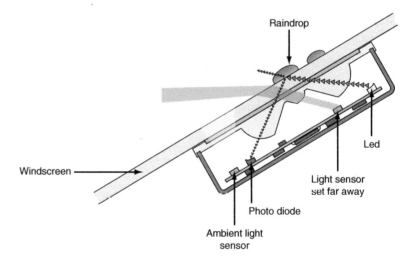

Speed sensitive wipers

Further technological advancement in wiper technology now allows the operation of the wipers to be linked to the vehicle speed. When the wipers are in operation the signals from the vehicle speed sensors along with the sensors on the wipers send signals to the ECU and from the data supplied the ECU can control the speed of the wipers. This also allows the wipers to be temporarily switched off when the vehicle is stationery, such as when it is at traffic lights.

Washer systems

A windscreen washer system is designed to spray fluid onto the windscreen and to work in conjunction with the wiper to keep the windscreen clean. The system usually has a permanent magnet motor which drives a centrifugal pump located in the washer fluid reservoir. Washer systems are activated by holding the washer switch and the pump spraying water onto the screen from the reservoir via two or more washer jets. One way valves are also fitted into the washer pipes. What is the purpose of these one way valves?

_____

_____

Washer systems are sometimes linked to the wiper circuit which will operate the wipers automatically once the washer switch is operated. The wipers will either make a few sweeps across the windscreen clearing the fluid or will operate in an intermittent mode for a short period of time.

On vehicles that have washers on the front and rear screens this can be operated by only having one washer pump with two outlets. When the front screen washer is operated the pump works in one direction blocking the outlet to the rear washer and when the rear washer is operated the pump works in reverse and the outlet to the front washer is blocked.

To diagnose faults on the washer system it is important to eliminate whether the fault lies within the electrical side, or the pipe and washer jet side of the system. The washer pump is usually fitted in the washer fluid reservoir and when operated makes a small noise. Operate the washer pump switch and check for a noise, if this is heard then check the pipes and washer jets for blockages. If the washer pump is suspected of being faulty list the checks that can be carried out to confirm this:

_____

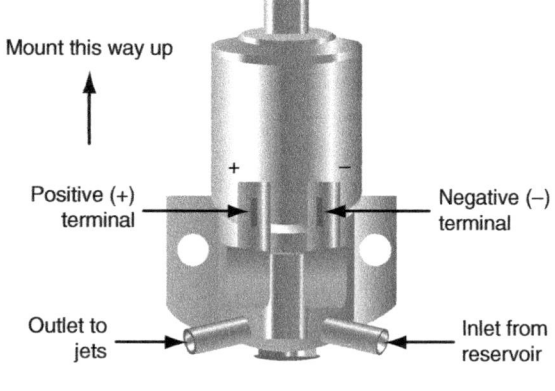

Mount this way up

Positive (+) terminal

Negative (−) terminal

Outlet to jets

Inlet from reservoir

_____

_____

_____

_____

Headlamp washer

Some vehicles are also fitted with a headlamp washer system to ensure the headlamps remain clean to give maximum visibility. The headlamp washer pump is usually a high powered pump that is actuated by a timer. The system can have its own switch or can be linked to the washer circuit. The need for headlamp washers is more relevant for vehicles with HID headlights because when the headlamp lens is dirty it disrupts the beam pattern and can dazzle oncoming traffic. Since 2010 manufacturers of vehicles fitted with HID headlights now have headlamp washers fitted as standard to prevent this.

ELECTRICALLY OPERATED WINDOWS AND MIRRORS

Windows

Side windows can be operated by electric motors connected to a gear linkage which moves the windows up or down, depending on the motor direction of rotation. A DC permanent magnet motor is used to power each window drive mechanism.

How is the direction of the motor reversed for up and down window operation?

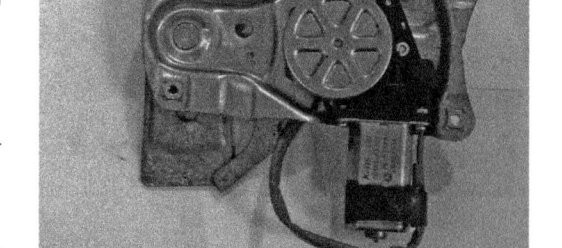

_____

_____

The wiring diagram on the right shows the main front/rear relays which are energised (fuse 16, wire GY) when the ignition is switched on.

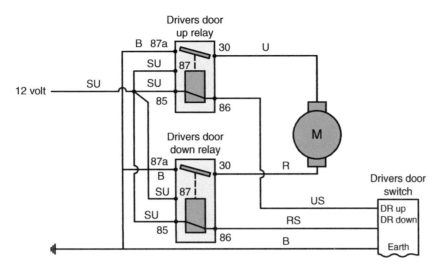

Describe the operation of the system during up and down operation of the driver's door windows.

Up _____

_____

_____

_____

_____

_____

Down _____

_____

_____

_____

_____

_____

_____

_____

_____

_____

_____

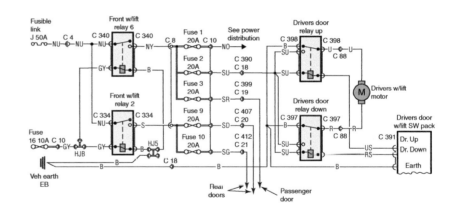

One touch operation

Some vehicles are fitted with a feature that allows the window to be fully opened with one touch. This function is operated by holding the down switch for more than 0.3 seconds then releasing it. The window can still be stopped at any time by operating the switch. This function is sometimes only incorporated into the driver's window.

Bounce back

Most electric windows incorporate a bounce back circuit which prevents the window from closing if something is in the way, such as fingers. On some systems if the window is closing and it hits something then the window will reverse and open. Other systems use infrared sensors to indicate if an obstacle is in the way.

Diagnosing faults on electrically operated windows

The electric windows use a motor and mechanical gear linkage system to operate. When diagnosing faults it is essential to identify whether the fault lies in the linkage or the electrical element of the system. Operate the electric window switch to see if the window moves or if it is moving very slowly. If the window is not moving disconnect the linkage from the motor and operate the switch again to see if the motor turns. If the motor turns the fault will most probably lie within the linkage system; this could be due to seizure or damage.

The first check on any electrical system should be to see if the fuse is still operational. If the fuse is ok then if the system has a relay fitted another quick check is to locate the relay, operate the system in question and see if the relay clicks, this will identify if the relay is energising or not. Another quick check is to operate the switch and using a test lamp probe check if the switch has a feed and if the feed is passing through the switch.

A customer complains the driver's electric window will not operate but all the remaining windows are working ok. Describe what steps should be taken to locate the fault.

- _____
- _____

_____

_____

_____

_____

_____

- _____

_____

- _____

_____

_____

_____

_____

_____

_____

Never test the bounce back operation by using your fingers or arm. If the system is faulty then you could cause serious injury to yourself.

Mirrors

Electrically adjusted rear view door mirrors are becoming more popular on modern vehicles. This allows the driver to adjust both door mirrors by an electrically operated set of switches usually located in the driver's door consul. Each mirror usually has two geared motors behind the mirror glass which allows for up/down and left/right movement of the mirrors.

How does the driver control the electrically adjustable mirrors?

_____

_____

_____

_____

The diagram below shows a typical power mirror circuit which is not linked to the memory power seats.

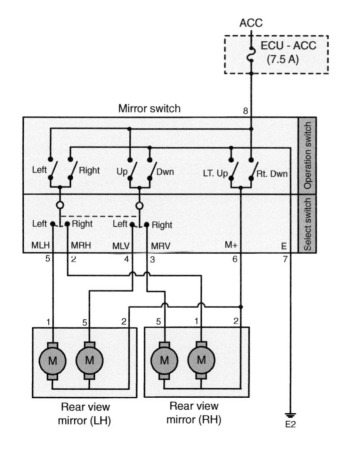

Heated rear screens are designed to clear the rear screen of ice and moisture as quickly as possible.

What components are fitted in the heated rear screen system to enable this to happen efficiently?

_____

_____

On some vehicles the heated rear screen will also operate a similar heating system in the rear view door mirrors allowing them to defrost at the same time as the rear screen.

How does the heated rear screen system operate?

_____

_____

The circuit can draw high current of around 10 to 15A. Due to this, the relay is usually on a timer and will automatically turn off after 10–15 minutes.

What is a quick way of testing that the relay is turning off the rear screen correctly?

_____

_____

_____

Research how a window heated rear screen heating element can be checked for damage.

_____

_____

_____

POWER SEATS

In older vehicles the driver adjusts the seat to the most comfortable driving position manually by using a number of levers and control knobs. On some modern vehicles this operation has been replaced by power seats using a number of switches which operate reversible permanent magnet motors. By using these motors it allows the driver to adjust the seat to the most comfortable driving position electrically.

Some power seats use three motors to give six-way movement of the seat, up/down, forward/rearwards and forward/rearward tilt. Others may incorporate a 4th motor to give eight-way movement which will also include the forwards/rearwards tilt of the seats backrest. The image on the right shows a seat which also incorporates a motor for the backrest adjustment.

Label the diagram on the right.

Memory seat function

Some more advanced vehicles have a memory seat option for automatically positioning the driver's seat to different positions. This function can allow up to three different seat positions to be memorised for different drivers. Some vehicles also have the seat memory position linked to the remote central locking key fob which allows the seat to be adjusted automatically once the door is unlocked.

Some vehicles have heated front windscreens to allow for quick defrosting. However, because the driver's vision cannot be obscured the standard demister element as used on heated rear screens cannot be used. Aircraft technology has been adapted to overcome this issue. Very thin wires or a microthin metallic coating is sandwiched between the layers of glass. The operation of the heated front windscreen is very similar to the rear screen and also draws a high current. The relay for the front heated screen will also have a timer incorporated into it.

The image below shows a typical wiring diagram for a heated rear screen with heated door mirrors incorporated in the circuit.

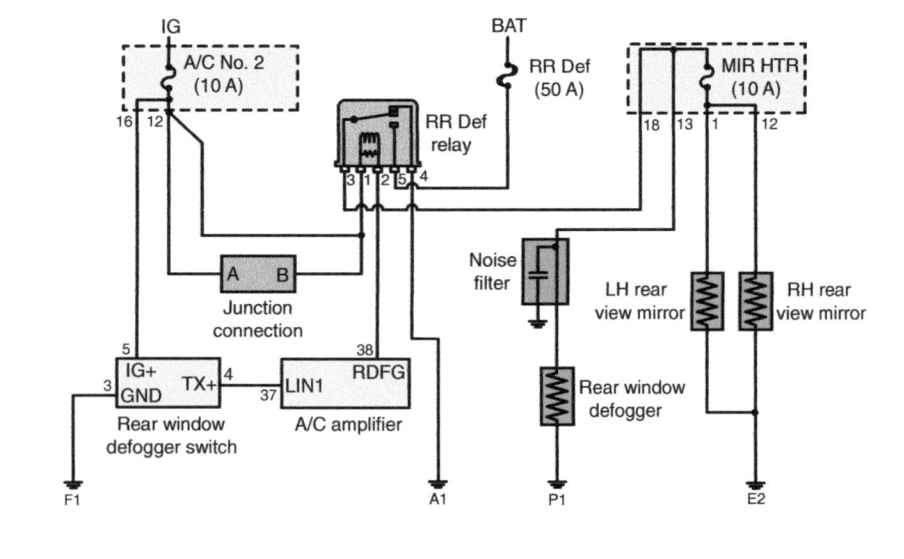

A central door locking system enables all doors, boot or tailgate to be unlocked or locked simultaneously when the key is turned in the door lock or the remote control operated. Electric motors or solenoids in each door and boot or tailgate are activated to operate the door locking mechanism.

Remote central door locking is controlled by a small hand-held transmitter and a receiver unit. When the system is operated by pressing the switch on the key or key fob a code is transmitted. The sensor in the receiver in the car picks up this code and if it is correct the relays are triggered and the door locks open.

The following shows a basic wiring diagram of a remote central locking system; with the aid of this diagram explain how the doors are unlocked:

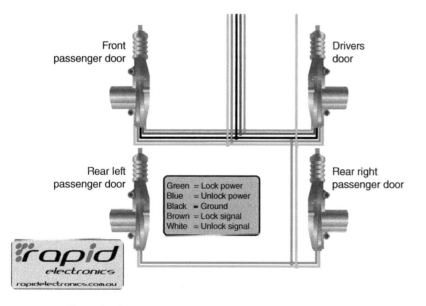

Front passenger door

Drivers door

Rear left passenger door

Rear right passenger door

Green = Lock power
Blue = Unlock power
Black = Ground
Brown = Lock signal
White = Unlock signal

Example of a remote central locking system wiring diagram

Diagnosing faults on the central locking system

One of the most common faults on the central locking system when the doors are not opening is that the battery in the remote key fob requires changing. This is a fairly easy, inexpensive and quick fault to repair. However, once the battery is changed the central locking still may not operate until it has been re-coded to the vehicle.

If the battery is not faulty then the technician will need to decide if the fault is in the electrical part of the system or the mechanical linkages part of the system. As with any electrical system a good starting point would be to check the fuse to see if it is serviceable. If the fuse is serviceable then disconnect the mechanical linkage from the central door locking actuators and press the remote locking key fob. If the actuators operate, then check all mechanical linkages for damage or seizure and replace where necessary. If the fuse has blown then there may be a fault in the wiring. Check all cables to the central locking door actuators for chaffing or damge causing a short circuit. Repair any damage found, replace the fuse and test the operation of the central locking system

If the linkages, fuse and wiring are OK and the actuators are not operating then there is an electrical fault on the system. As the fuse and key fob have been deemed serviceable, disconnect the electrical connector on the driver's door actuator. Using a test lamp probe find a suitable earth and connect the cable then probe the electrical connector whilst pressing the key fob. The light should illuminate if there is a feed to the actuator. If the light does not iluminate check the wires for continuity using a multi-meter.

If there is a feed but the actuator is still not operating state how you would check the actuator for serviceabllity.

TIP

When the battery is changed on a remote locking key it may require re-coding to the vehicle. Follow the manufacturer's guidelines to ensure the key operates in the correct manner.

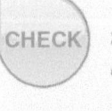

Some remote units and their receiver change access codes each time the remote is operated to give enhanced security.

Smart keys

Many vehicles no longer require the use of a traditional key to gain access to the vehicle or to start/stop the engine. Instead of using a traditional key a smart key is used which relies on a two-way communication system between the smart key and various antennas situated around the vehicle.

What is located in the smart key to allow for communication with the antennas around the vehicle?

In early versions of this technology a button was pressed on the smart key to gain access to the vehicle and then the smart key was inserted into a slot in the vehicle dash and a button pressed to start the engine.

By using the two-way communication between the smart key and the vehicle antennas this will only allow access to the vehicle and the engine to start if the signal transmitted by the smart key is received and matched with the data stored in the vehicle system.

Some more advanced smart keys do not need to be inserted into the vehicle nor have any buttons pressed to open the vehicle. This type of key has a radio pulse generator inside which emits a radio signal. This signal is picked up by several receiver antennas on the vehicle when the key is within a certain zone around the vehicle. The code is verified as correct and then the door locking system operates.

HEATING, COOLING AND AIRCONDITIONING

Passenger comfort is an important aspect of modern vehicle design. One way of ensuring this comfort is to provide a suitable ventilation, heating or air conditioning system.

How does the ventilation system work?

This system works extremely well when the vehicle is moving; however, when the vehicle is stationery the steady flow of air stops. What system is used to keep the air circulating when the vehicle is stationary?

When the vehicle is moving, the supply of outside air is continuous and this could cause a problem by over-pressurising the passenger compartment. To alleviate this issue the passenger compartment is fitted with exit vents and/or a pressure relief valve. These can be found in the door seals, door posts or under the rear seats. The image opposite shows a typical 'flow-through' ventilation system.

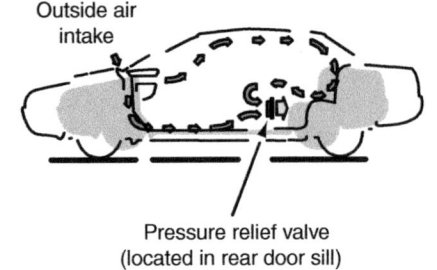

Outside air intake

Pressure relief valve (located in rear door sill)

Using your own knowledge and skills along with manufacturer's data, inspect a vehicle and see if you can find the pressure relief valve and exit vents.

Heating

The vehicle heater is made from an airbox and heater matrix. The heater matrix is part of the vehicle cooling system and has a steady flow of coolant through it. Air is drawn into the heater via the ventilation system and is then distributed in the passenger compartment via vent ducts at different levels. The vent ducts usually allow for windscreen demisting, face level vents and floor level vents both in the front and rear passenger compartments.

When diagnosing faults on the heater system it is essential to consider what is happening to the engine cooling system as well. For example, if the heater is blowing out cold air but the engine temperature is hot then the fault will lie in the heating system from its connection to the engine cooling system, such as the connecting pipes, vents or heater matrix. However, if the heater is blowing cold air and the engine coolant temperature is not rising this is more likely to be a fault on the engine cooling system, such as a faulty thermostat.

The following image shows how the heater is connected to the vehicle coolant system.

Label the diagram with the following:

heater hoses	**thermostat**	**expansion tank**	**heater control valve**
heater core	**radiator**	**water pump**	

The level of heat that is dissipated into the passenger compartment is controlled by two switches. One switch controls the fan speed and the other switch controls how much of the air flows over the heater matrix before entering the passenger compartment. The air that flows over the heater matrix transfers the heat from the coolant to the air and then into the passenger compartment – this is a form of heat exchanger.

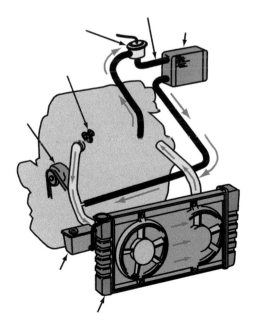

When working on any heating system ensure the coolant has had time to cool down before removing any pipe or you could get hot coolant over you causing burns.

Passenger comfort is important to vehicle manufacturers and over the years advances have been made to heating systems. Modern heating systems can be programmed to have different temperatures for different areas of the passenger compartment, for example the driver and front seat passenger can have different temperatures and also the front of the vehicle and the rear of the vehicle can have different temperatures. This type of system is known as climate control. This picture opposite shows a modern climate control system.

Airconditioning

Over recent years manufacturers have taken even more interest in passenger comfort and more vehicles are now produced with airconditioning as standard, which gives a different form of heating and cooling. The air-conditioning system has advantages over the normal heating system such as maintaining a comfortable temperature in the passenger compartment in hot weather, dehumidifying the air in hot weather and dehumidifying the air in cold weather which prevents windows from misting up. However, there are a number of disadvantages to air-conditioning systems over conventional heating systems.

List three disadvantages:

1 _____

2 _____

3 _____

Safety precautions

Air-conditioning systems are filled with refrigerants under high pressure. The 'Montreal Protocol' is a treaty that was signed to reduce the effect of CFC gases on the ozone layer. Since 1993 the refrigerant R12 (Freon CFC) has been replaced by the refrigerant R134a which is a HFC gas and less damaging to the ozone layer. What do CFC and HFC stand for?

CFC: _____

HFC: _____

When working on the air-conditioning system the fittings must not be loosened or components removed until the refrigerant has been correctly discharged, it is illegal to discharge refrigerant gases into the atmosphere. New legislation (EC 842-2006) means that since July 2010 all individuals working on mobile air-conditioning systems must have achieved, as a minimum requirement, a refrigerant handling qualification.

 Research and give a brief explanation of the main considerations of the Montreal Protocol and the main legislative requirements of EC842-2006 (the websites below may be useful).

 http://ozone.unep.org/pdfs/Montreal-Protocol2000.pdf

https://www.gov.uk/environmental-regulations

Working with refrigerants presents its own hazards; give a brief example of how the refrigerant is dangerous next to each item listed:

Skin contact: _____

Suffocation: _____

_____

_____

Flammability: _____

_____

To prevent skin contact certain PPE should always be worn when working on air-conditioning systems; list the four main items of PPE that must be worn:

1 _____

2 _____

3 _____

4 _____

Refrigerant is stored in pressurised containers like the one shown in the picture; these cylinders are usually colour coded for easy recognition.

Basic refrigerant circuit

On the diagram below, show the direction of refrigerant flow and indicate where changes of state take place. Complete the text boxes by explaining what is happening to the refrigerant in each component:

Passenger compartment

Evaporator

Evaporator:

Low side tap

Expansion valve (or orifice tube)

Expansion valve:

Compressor

Compressor:

High side tap

Condenser

Condenser:

Receiver/ Dryer:

Front of car

Receiver dryer

Major components

Compressor

The compressor is a key element to the air-conditioning system. There are a number of different types of compressor currently available including rotary vane, wobble plate and screw type; however, they all serve the same purposes and that is to draw the low pressure refrigerant gas from the evaporator, increase the pressure and temperature of the gas and circulate the high pressure, high temperature refrigerant around the system.

The compressor is driven by a belt from the engine which means the pulley is always turning; however, the compressor only operates once the electromagnetic compressor clutch is engaged. When the airconditioning is switched on a 'live' feed is sent to the clutch which then activates an electromagnetic coil to engage the clutch with the pulley. Once the airconditioning is turned off the 'live' feed is removed from the clutch and the electromagnetic coil deactivates and the clutch in turn disengages.

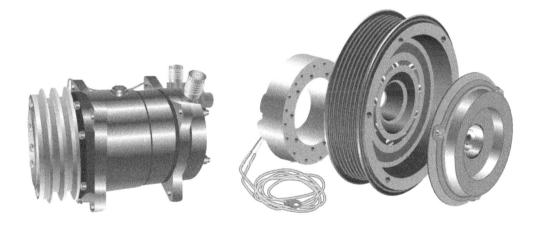

A vehicle has come into the workshop with a suspected broken air conditioning compressor clutch, state what checks to carry out to confirm the fault.

_____

_____

_____

_____

_____

_____

_____

Condenser

The condenser is situated at the front of the vehicle and looks like a separate radiator. This will usually be situated in front of the vehicle radiator to aid its operation. The condenser receives the high pressure, high temperature refrigerant gas from the compressor, the cool air passing over the condenser causes the refrigerant to cool down and change into a liquid state, therefore acting as an heat exchanger. The cool air can be from ram air when the vehicle is moving or by a fan when the vehicle is stationery.

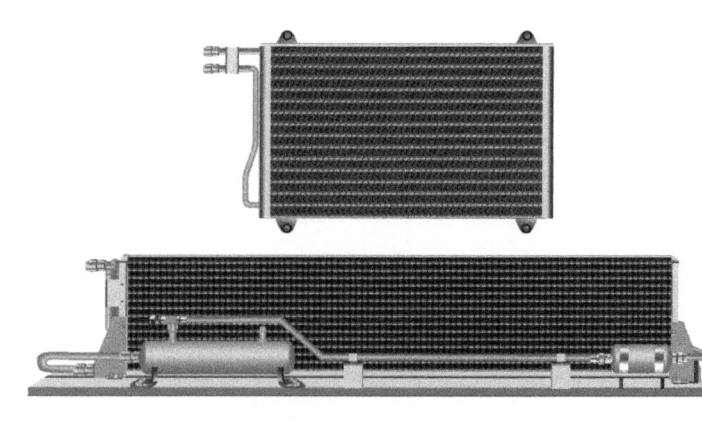

Receiver/dryer

The receiver/dryer unit is sometimes called the filter/dryer and acts as a small reservoir, filter and moisture removal system. The receiver/dryer is located after the condenser in the system but can be attached to the side of the condenser.

Explain the operation of the receiver/dryer: _____

_____

_____

_____

Like any filter the receiver/dryer will reduce in efficiency over time and will require replacing according to manufacturer's recommendations.

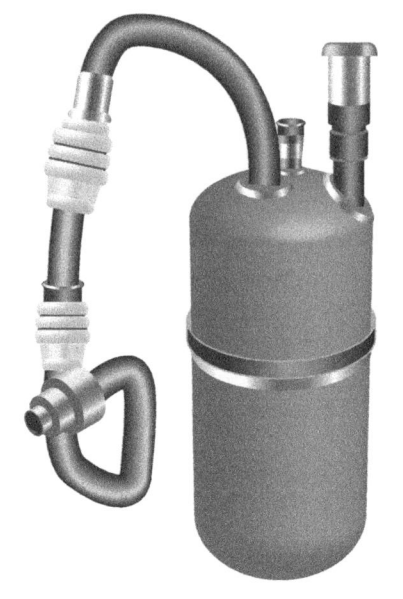

Expansion valve

Liquid refrigerant is passed to the expansion valve from the receiver/dryer. The expansion valve has a small orifice inside through which the liquid refrigerant must pass. The expansion valve delivers the required refrigerant dependent on the evaporator temperature and system demand.

The orifice causes a restriction of the liquid refrigerant which causes the pressure to drop and in turn the temperature drops. The refrigerant that leaves the expansion valve is a low pressure mixture of liquid and gas refrigerant.

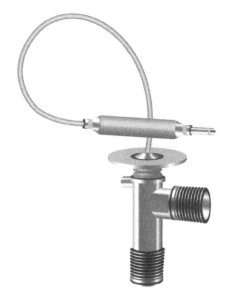

Evaporator

The evaporator acts like a heat exchanger; it has a winding tube construction surrounded by fins. It is designed to absorb the maximum amount of heat in the smallest possible space. The warm air flows through the fins and over the tubes containing the refrigerant.

Explain what happens to the refrigerant as it passes through the evaporator.

_____

_____

_____

_____

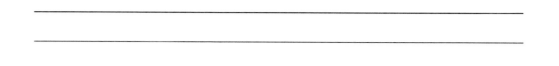

Oils

Various components in the air-conditioning system require lubrication such as the compressor. The oil is circulated around the system with the refrigerant.

Service/charging ports

In order for the air-conditioning system to be serviced charging ports are located on the system and can be fitted to hoses, pipes and receiver/dryers. The high pressure and low pressure charging ports have different size ports for differentiation between the two. To help prevent any leakage from the ports a plastic cap with a rubber seal is used to close the charging ports when not connected to charging systems.

Servicing equipment

When servicing and maintaining the air-conditioning system different equipment can be used. A recovery/recycling unit, evacuation unit, charging station and manifold gauges are the most common items. However, with advances in technology, all of these items are now incorporated into one machine. A modern all-in-one unit can be seen here.

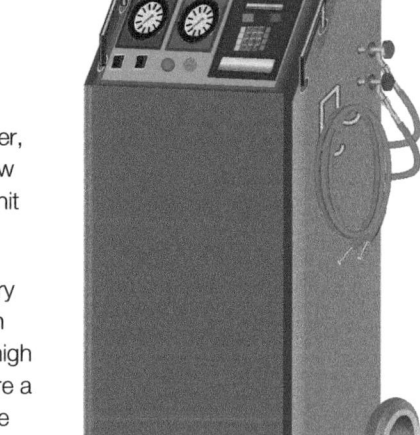

The manifold gauges are not only used for the recovery and charging of the air-conditioning system but also in fault diagnosis. The readings on the gauges for both high and low pressure sides of the system can identify where a particular problem may be (this will be covered in more depth later in the chapter). The following is a guide on how to recover, evacuate and recharge an air-conditioning system.

Recovery

Turn the all in one machine on and open both the high and low pressure manifold gauge valves. Connect the blue hose to the low pressure service valve and the red hose to the high pressure service valve; once connected open both the service valves. Select the **recovery** setting on the machine and ensure any readout is reading zero and start the machine. Wait until the recovery is complete (some machines will have a light to show this whilst others will have a digital display). Open the oil drain valve, measure and record the drained oil.

Evacuate

Following the recovery keep both service valves open and choose the **evacuate** setting on the machine; set the recommended timing on the machine timer and start the machine. When the timer expires close both service valves and check the manifold gauges to confirm the vacuum. Open the oil injection valves to inject the oil into the air-conditioning system as required.

Charge

Open the low side service valve and keep the high side service valve closed. Choose the **charge** setting on the machine, set the machine to the desired weight of charge required (this will be

manufacture and system specific) and then start the machine. The machine will stop once the desired charge has been completed. Close the low side valve.

Some manufacturers have an easy to use guide on the machine like the one shown here:

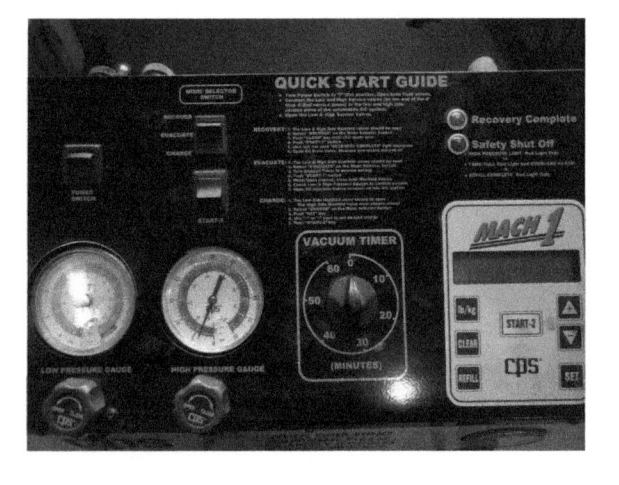

 Always ensure when using any air-conditioning machine that the manufacturer's instructions are followed correctly.

 Under supervision of your tutor/assessor carry out a recovery, evacuation and recharge of a vehicle air-conditioning system.

Leak detecting

Leak detecting is a very important part of the service procedure as a low charge of refrigerant will cause system damage; also if there is a leak the lubricatiing oil can escape causing catastrophic component failure. There are three main types of leak detection: visual, electronic and dye. A quick visual check of the connections may show some leaks as there may be signs of oil or encrusted dust; however, this is not a definitive check and more technical methods are more accurate.

Electronic leak testing

The electronic leak detectors can operate in a number of ways but one of the most popular is a detector that emits a ticking sound when turned on, this sound turns into a high pitched noise when a leak is detected. The detector has a sensing tip that should be moved around the components and connections about 5mm from them. If the sensing tip touches any components it can give a false reading and cause damage to the detector.

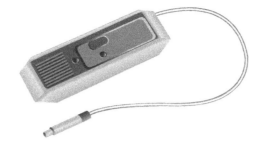

Dye detector

A fluorescent dye is injected into the air-conditioning system and circulated with the refrigerant. A special ultraviolet lamp is then used to detect the leaks by passing it over the components and connections, the fluorescent dye glows brightly under this lamp. This type of detection is extremely good as it can pinpoint small leaks.

 Carry out a leak detection test on a functional air-conditioning system

Performance test

Prior to carrying out a performance test on a vehicle's air-conditioning system a number of preliminary checks should be completed.

List the preliminary checks that should be completed:

- _____
- _____
- _____
- _____
- _____
- _____

Once these preliminary checks are satisfied a full performance test can be carried out.

Research and list the procedure for a typical air-conditioning performance test:

1 _____
2 _____
3 _____
4 _____
5 _____
6 _____
_____
7 _____

8 _____

9 _____

_____

Diagnosing air conditioning faults

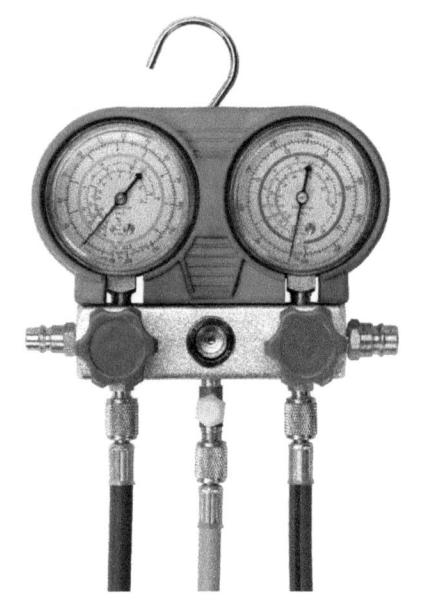

When diagnosing faults on the air-conditioning system the technician must be confident of reading the manifold gauges correctly. These gauges are an essential item for diagnosing faults and determining where in the system the fault lies. Prior to carrying out any diagnostic work the gauges must be set to zero. Once zeroed, connect the manifold gauges to the system and monitor the readings. The manifold gauge readings are dependent upon ambient temperature, however, for a system that is operating normally the readings should be between 1.6–2.0 bar (24–30psi) for the low pressure side and 10–15 bar (150–220psi) for the high pressure side.

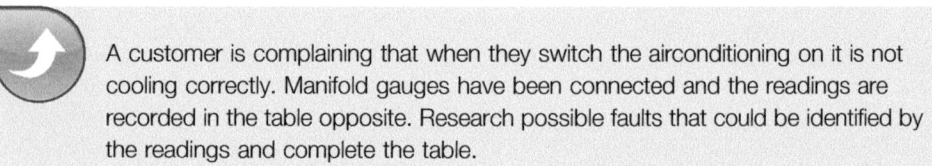

A customer is complaining that when they switch the airconditioning on it is not cooling correctly. Manifold gauges have been connected and the readings are recorded in the table opposite. Research possible faults that could be identified by the readings and complete the table.

Low pressure reading	High pressure reading	Possible causes
Normal	Normal	• _____
		• _____
		• _____
Normal or low	Low	• _____
		• _____
		• _____
		• _____
		• _____
High or normal	High	• _____
		• _____
		• _____
		• _____
		• _____
Equal to high pressure	Low	• _____
		• _____
		• _____

Low pressure reading	High pressure reading	Possible causes
High	Normal or low	• _____
		_____
		• _____
		• _____
		• _____
		• _____
Low	High or normal	• _____
		• _____
		_____
		• _____

3 **The relay in a heated rear screen has a timer built in, this is to**

  a) Turn the heater off after a pre-determined period of time [ ]

  b) Turn the heater off when the window has fully defrosted [ ]

  c) To keep the heater on as long as the ignition is on [ ]

4 **An airconditioning compressor clutch will usually be of what type?**

  a) Dog clutch [ ]

  b) Friction clutch [ ]

  c) Electromagnetic clutch [ ]

5 **How many brushes are fitted to a two-speed wiper motor?**

  a) 2 [ ]

  b) 3 [ ]

  c) 4 [ ]

6 **A power seat with six way adjustment would usually have a minimum of how many motors?**

  a) 3 [ ]

  b) 6 [ ]

  c) 4 [ ]

7 **A one way valve is fitted into the washer pipes for what reason?**

  a) To allow fluid to return to the reservoir [ ]

  b) To prevent siphoning of fluid back to the reservoir [ ]

  c) To allow fluid to pass out of one jet only [ ]

Multiple choice questions

Choose the correct answer from a), b) or c) and place a tick [✓] after your answer.

1 **What is the name of the refrigerant used in modern air-conditioning systems?**

  a) R134a [ ]

  b) R135a [ ]

  c) R12 [ ]

2 **In a standard heating system; which of the following is NOT part of the system?**

  a) Heater matrix [ ]

  b) Air vents [ ]

  c) Receiver/drier [ ]

PART 7
BODY REPAIR

USE THIS SPACE FOR LEARNER NOTES

_____

_____

_____

_____

_____

_____

SECTION 1
Mechanical, electrical and trim components 305

1 Body repair health and safety 306
2 Body repair 307
3 Vehicle manufacture 308
4 Final adjustments to body repair 312
5 Common electrical components 313
6 Adding modifications to vehicles 319
7 Removing and replacing trim components 321
8 Review 328
9 Multiple choice questions 329

SECTION 1

Mechanical, electrical and trim components

USE THIS SPACE FOR LEARNER NOTES

_____

_____

_____

_____

Learning objectives

After studying this section you should be able to:

● Understand mechanical and electrical components and safety.
● Describe methods of adjusting components.
● Understand vehicle manufacture and modifications.

Key terms

Incandescent bulb   A bulb that uses a wire or filament which heats up to give off light.
Corrosion   A term used for rust.
ABS   Anti-lock braking system.
Vinyl wrap   Vehicle covering used to advertise or change colour.

WWW   http://ask.cars.com/2008/08/xenon-hid-headl.html

http://www.tradingdirect.co.uk/buyingguides/parking_sensors/work.aspx

BODY REPAIR HEALTH AND SAFETY

As a Level 3 student you should have a good knowledge of health and safety and use of tools, here are a few items of safety that may be new to you.

Orange wiring

The wiring harness cover and connector on the above vehicle are orange, what may this indicate?

_____

_____

 Always follow manufacturer's instructions when working on high voltage systems.

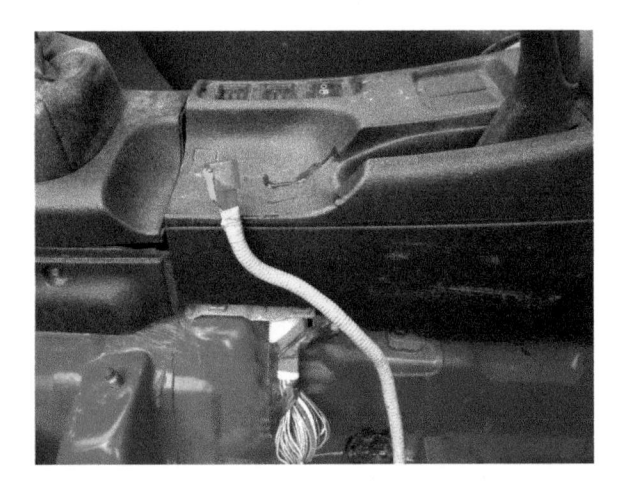

Wiring harness

The wiring and harness in the image above has a yellow cover. What does the yellow wiring harness cover indicate?

_____

_____

 **TIP** Remember: you must never join accessories into this wiring harness such as stereos, tow-bar wiring or Bluetooth kits etc.

 You must never check these circuits with test lamps, multi-meters or power probes as you can accidently detonate airbag systems.

You may be expected to remove and replace air conditioning parts during the repair of vehicles after accidents. You can remove and refit any parts of the air conditioning system but you will need to be trained and certified to complete which task?

_____

_____

The below picture shows an air conditioning condenser, discuss with your classmates the procedure in order to replace this part. Make notes below.

Air conditioning

_____

_____

_____

 Research colleges in your area that will complete an air conditioning gas handling course with you.

Batteries can be dangerous if not handled correctly. Always ensure that you disconnect them and connect them in the correct order to prevent short circuits.

What is the correct sequence for disconnecting a battery?

1 _____

2 _____

3 _____

4 _____

If a bonnet has been damaged and folded due to the crumple zones, there is a danger involved with the battery, what is the danger?

_____

_____

_____

 Disconnect or remove the battery on the vehicle whilst it's standing waiting for repair work to be authorised by insurance companies.

BODY REPAIR

Body repair is as complicated a subject as maintenance and repair and therefore MET (mechanical, electrical and trim) technicians have a very responsible job. It may be that you start to work in a garage and are asked to remove components ready for complicated body repair or jig work, often this job role is known as a 'stripper and fitter'. The main responsibility is returning the vehicle to the customer in as near-to-perfect condition as you can.

Scenario

The weekend mechanic that attempted to restore the vehicle in the image below has made a huge mistake; do you know what has been done incorrectly?

_____

_____

_____

_____

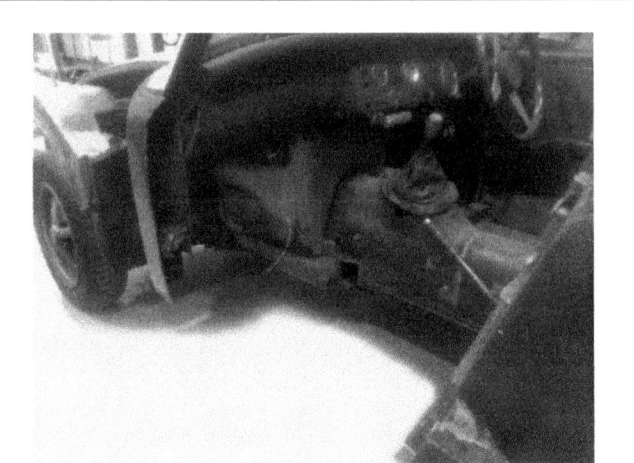

Restoration project

 If you are in doubt about anything then always ask your supervisor. Mistakes can be very costly to repair, not just with structural panels but mechanical parts too.

VEHICLE MANUFACTURE

Vehicle body design

Vehicle body design has changed quite dramatically in recent years, if you compare a modern car to a car of 10 or 15 years ago you will notice that now many vehicles are much bigger and there is a lot more emphasis on safety.

 In small groups list the safety features on new cars compared to those of 10–15 years ago.

_____

_____

_____

_____

The vehicle on page 308 has crumple zones built into it to protect the occupants in an accident, how does a crumple zone work?

_____

_____

_____

_____

_____

The image above shows a side impact bar found in vehicle doors. As well as protecting car occupants in a side impact, what else does the side impact bar do?

_____

_____

■ Mild steel
■ High-strength steel
■ Extra-high-strength steel
■ Ultra-high-strength steel
■ Hydroformed aluminium

■ Mild steel
■ High-strength steel
■ Extra-high-strength steel
■ Ultra-high-strength steel

Vehicle body structure

Modern vehicle manufacture uses a range of metals to ensure that the occupants are kept as safe as possible. Different strength steels may be used for different areas of the vehicle but while high strength steels offer lots of protection what are the disadvantages of them?

1 _____

2 _____

_____

Study the picture of the vehicle body structure above. Using the list below, add numbers to the diagram to show the corresponding parts.

1 **Front bumper reinforcement**   5 **The B Post**

2 **Front crumple zone**   6 **The C post**

3 **Strut tower**   7 **The sill**

4 **A post**   8 **The bulkhead**

The shape of some vehicle components can be quite complicated, modern manufacturing techniques often involve hydro-forming of the metal shapes.

Research the process of hydro-forming, find out how it is used to form the shapes in metal.

You may wonder how modern manufacturers manage to achieve such a good paint finish on their vehicles, and you may have noticed that vehicles do not corrode as badly as they used to. What do manufacturers do to car bodies to prevent corrosion?

_____

_____

_____

_____

Primer is the first layer of paint used on vehicles. This may involve dipping the vehicle shell into a huge bath which will allow the primer to get into parts that spray painting could not reach. It is made to fill tiny scratches and imperfections in the body and give the subsequent colour coats something to stick to. Many manufacturers have begun to use coloured primers to hide future damage from stone chips.

Before painting, some manufacturers brush the vehicle with ostrich feathers which have an electrostatic charge passed through them, what is the purpose of this process?

_____

_____

_____

www **http://www.bbc.co.uk/wear/content/image_galleries/nissan_gallery .shtml?25**

The basecoat, or the colour layer, is next in the process, then clear-coat paint or lacquer is generally applied over the final colour layers. Clear-coat paint is a layer of paint without colour pigment added – it is simply clear paint.

What is the main purpose of the clear lacquer?

_____

_____

The car body may be charged with electricity to allow the paint to be 'attracted' to the body. This allows less waste and better coverage of the vehicle. Research and describe how this operation works.

_____

_____

_____

_____

_____

Production line paint

What is another advantage of this electrostatic process?

_____

_____

Modern paint techniques allow for the use of water-based paint to be used as a basecoat but why does the lacquer still need to be solvent-based?

_____

_____

Carrying out checks to give customers an estimate

If a vehicle comes into the garage after an accident you may be expected to provide an estimate of the cost to repair.

What is the difference between an estimate and a quote?

_____

_____

Why does a body repair centre prefer to offer an estimate rather than a quote?

_____

_____

_____

_____

Accident damage

What may be considered as mechanical components?

Body repair isn't just about repairing dents and painting panels; often an accident will involve damage to mechanical components. Which components will need to be replaced depends on the severity of the accident. Mechanical components are not just limited to engines and gearboxes, a MET technician should have a good knowledge of most vehicle systems and will commonly need to replace steering and suspension components which will need to be aligned once fitted.

Suspension and steering should be adjusted correctly

Setting up the steering and suspension

Once you have completed the removal and replacement of parts it is important that the vehicle is 'aligned' properly so that it drives correctly on the road. You may need to complete a four-wheel alignment procedure and make adjustments as necessary.

Complete the drawings opposite and on page 312 to show the steering angle being checked.

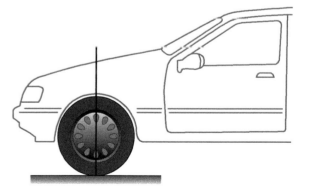

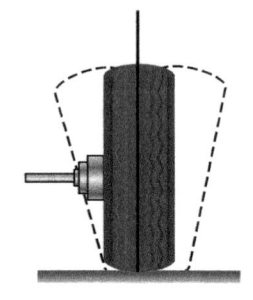
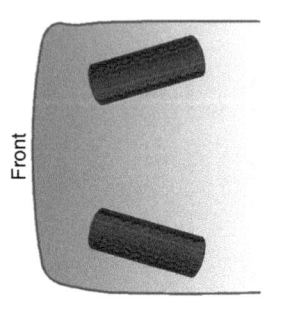
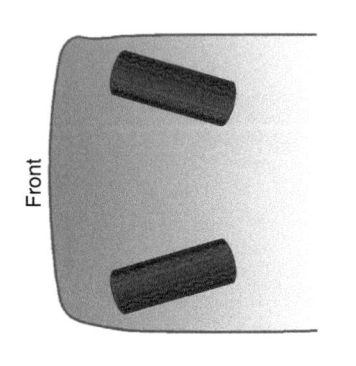

What could have caused this damage?

_____

Check link

CHECK Always set steering and suspension geometry to manufacturer's specification. More detail can be found in the chassis system of this book.

The check link as shown in the diagram above is inside the door, you can see that it should be bolted into position.

FINAL ADJUSTMENTS TO BODY REPAIR

TIP The check link should be lightly oiled to allow a smooth and quiet operation.

Before returning the vehicle to the customer a quality check may be required, this will ensure the vehicle is fully working and that it looks good.

What has happened in the figure opposite?

_____

_____

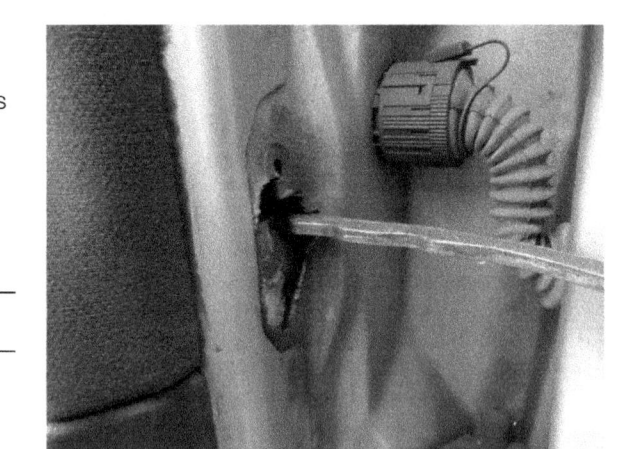

Check link

Bonnets and doors may need adjustment to ensure they close correctly and look good.

When simply removing and refitting body parts often you can refit them in the position that they came from. The photos opposite show some bonnet hinge bolts, the dirt around the hinge and a mark in the paint will create a perfect mark to refit in the position that they came from.

Bonnet hinge

Rubber stopper

If you are fitting a new front panel adjustment can be made by adjusting the bonnet catch or cable, and then further minor adjustment can be made by twisting the rubber stoppers at either side of the bonnet, as shown in the image above.

Using the pictures explain how a vehicle door is adjusted to close correctly.

_____

_____

_____

_____

Door striker plate

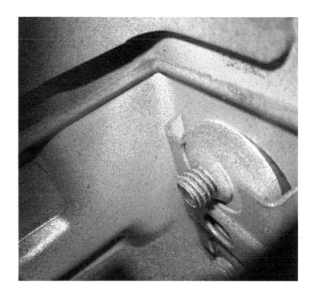

Inner side of striker plate

The image above shows the inside of a door striker plate, movement is allowed for adjustment but the inner plate should not be allowed to fall out of place if the bolts are fully removed.

COMMON ELECTRICAL COMPONENTS

The headlight shown in the image on page 314 is an HID light (high intensity discharge lamp). These are often referred to as xenon lamps (pronounced zenon) because of the xenon inert gas inside the bulb.

What is the difference between a xenon bulb and a traditional incandescent headlamp bulb?

_____

_____

Xenon headlight

Xenon headlamps can be recognised by the blue tint that they give off, they are brighter than traditional headlamp bulbs and you should ensure they are aligned correctly.

What do most manufacturers include on a vehicle to help prevent oncoming drivers being dazzled by xenon headlamps?

_____

_____

_____

_____

_____

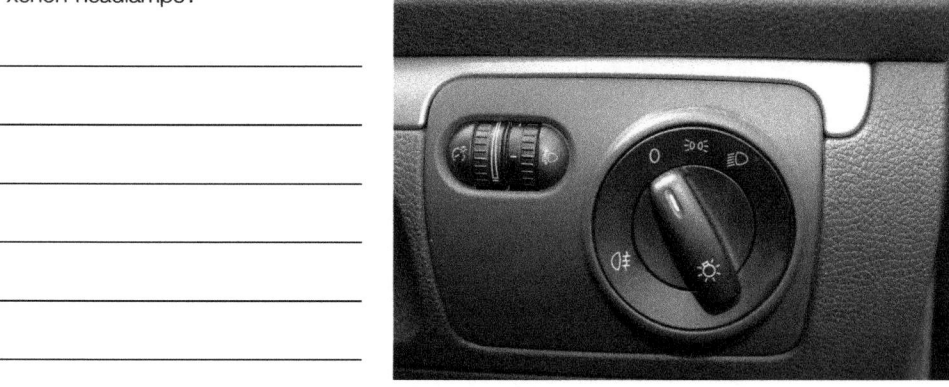

Headlamp switch

 http://ask.cars.com/2008/08/xenon-hid-headl.html

What may be considered as dangers of xenon bulbs?

_____

_____

_____

What should you consider when replacing a xenon bulb?

• _____

_____

• _____

_____

• _____

_____

• _____

CHECK   Always follow manufacturer's instructions when replacing a xenon bulb.

Adaptive lighting has been developed which uses a steering angle sensor on the steering wheel to shine the headlamps around corners which is proportional to the steering effort made by the driver.

Xenon headlamp bulb

Research adaptive headlight systems and find out how they work. Complete this task and share your findings with your group.

Airbags

Using the images below explain the correct sequence to remove and refit an airbag unit.

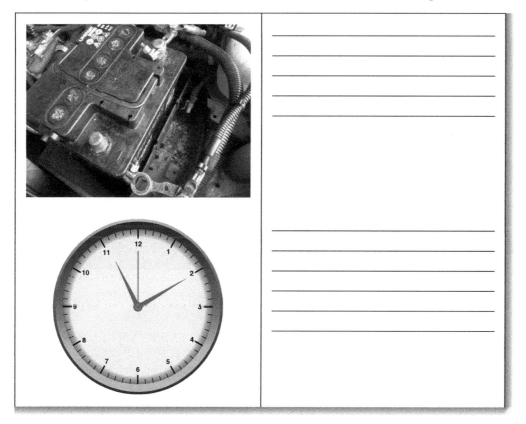

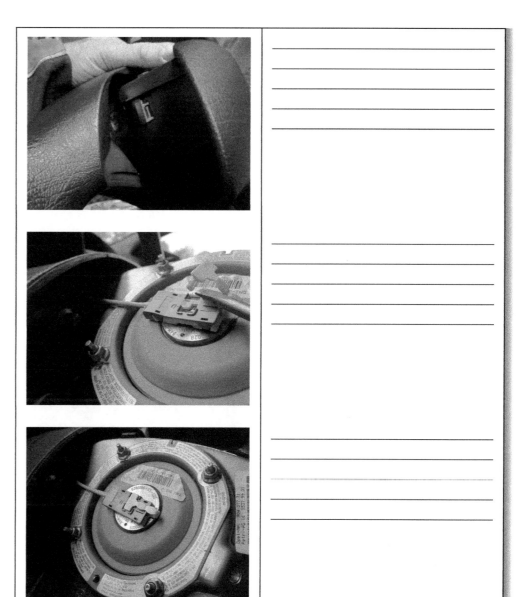

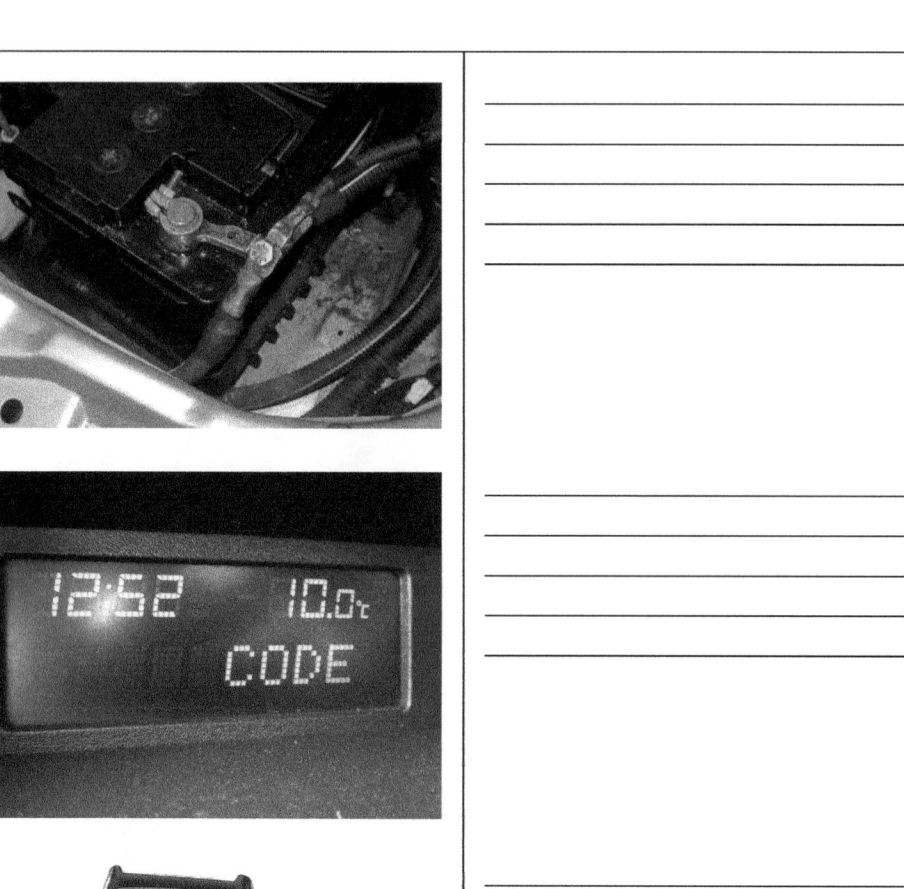

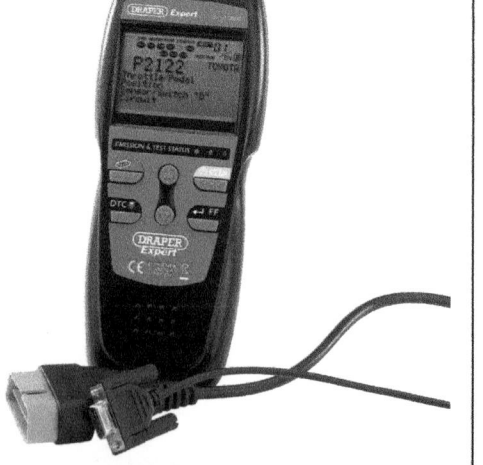

Impact sensors or crash sensors as they are often called are usually located somewhere under the bonnet at the front of the vehicle, perhaps behind the headlamp.

Some manufacturers fit sensors that automatically reset themselves after an accident, others need to be replaced. How could you tell if the sensor is the type that resets?

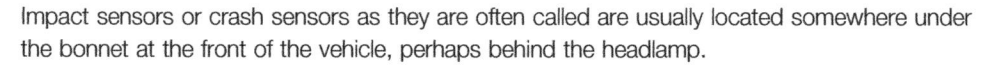

Research different types of crash sensors so that you can recognise them on different vehicles.

You may want to consider that the sensor could be damaged internally and that it is of paramount importance that it works, so if you are in doubt it should be replaced.

 Disconnect the electrical system before you remove and replace airbag components.

Sensor

A seat belt pre-tensioner may work in conjunction with the airbag. What is a seat belt pre-tensioner?

_____

_____

_____

Once a pre-tensioner system has activated the airbag light will be illuminated to warn that it needs replacing.

Some airbags are known as smart airbags which may inflate in two stages.

Wiper motors and linkages

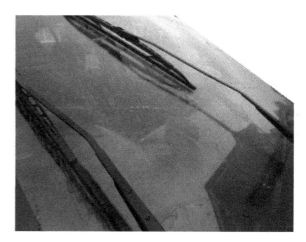

Wiper arms

Wiper motor

If you are involved in the replacement of windscreens or trims you may be expected to remove and replace a wiper motor and linkage. What is the biggest danger of working with wiper motors and linkages?

_____

_____

If the system is not working how can it be tested?

_____

_____

_____

TIP — Some wiper motors use their mounting bracket as an earth so check for good clean connections.

Once you have refitted the motor and linkage how should you check the system?

_____

_____

_____

Finally check that the wiper blades are not split and that they do not smear the windscreen in operation.

Testing a wiper motor

Anti-lock braking systems (ABS)

When replacing hubs and suspension parts you may come across **ABS** sensors that have been damaged and need replacing. The problem you could face here is that the ABS may have been faulty before the accident and could have even been a contributory factor of the accident. How can you test the ABS system to locate possible faults?

_____

_____

You could complete further checks with multi-meters or oscilloscopes, more information can be found in Part 4 (brakes section) of this book.

Once you have completed the job how can you be certain that the ABS has been repaired correctly and working as it should?

_____

_____

Remember the ABS may be connected to other systems so also check these systems for operation. Which other systems work through the ABS?

- _____
- _____
- _____
- _____
- _____

ABS sensor

ADDING MODIFICATIONS TO VEHICLES

Carrying out modifications is common on vehicles and one modification you may be expected to complete is the fitting of rear parking sensors.

Research rear parking sensors and then explain how the two common types of sensor work below.

Ultrasonic sensor

_____
_____
_____
_____

Electromagnetic sensors

_____
_____
_____
_____

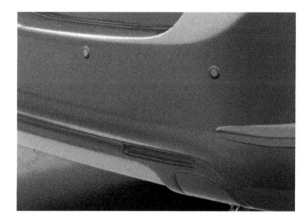

Rear parking sensors

TIP

It should be noted that parking sensors are only a guide and the driver is still responsible to use his sight and judgment when reversing. It would be worth discussing this with customers when you show them how to operate the system.

www http://www.tradingdirect.co.uk/buyingguides/parking_sensors/work.aspx

Spoilers and body kits

Many people add spoilers and body kits to their vehicles in order to enhance them, often they think that a spoiler the size of an aeroplane tail fastened to their boot lid will enhance performance of the vehicle. Unfortunately, for the speeds that we can legally drive, a spoiler has little effect apart from being aesthetically pleasing. The desired look can have negative effects, because what one person likes another may not which can affect re-sale values of cars.

Spoiler kits

When it comes to motorsport and high speed driving, spoilers are essential.

 In small groups make a list of modifications for safety and modifications for styling.

What is the purpose of the chin spoiler at the bottom of the lower bumper on the car in the image above (you may know this as a splitter or skirt).

_____

The car in the image opposite also has a rear wing spoiler mounted on the boot lid, what is the purpose of this spoiler?

_____

_____

Vinyl wrapping

Vinyl wrapping of vehicles is becoming very popular. It involves applying vinyl sheets to a vehicle. What are the main advantages of a **vinyl wrap**?

● _____

● _____

● _____

_____

● _____

Are there any disadvantages to a vinyl wrapped car?

● _____

_____

_____

● _____

TIP

You should check the current legal status of your vehicle after a wrap, if the colour has completely changed you may need to inform the DVLA (Driver and Vehicle Licensing Agency) and insurance companies.

The vehicle should be fully prepared and clean before adding the wrap.

 www.wrapyourcurves.co.uk

REMOVING AND REPLACING TRIM COMPONENTS

Removing trim components can be difficult, especially with customers wanting to personalise their vehicle with vinyl wraps or painted trims. If you scratch these on removal the customer will be sure to return with a complaint.

Research companies that complete 'smart' repairs and discuss in groups the advantages of 'smart' repair companies.

1 Have a look at the task first and check the tools and equipment required. Look for hidden screws and clips.

Take your time and examine the trims first, look for clips and screws to remove them.

If you are unsure of how trims come off what can you do?

_____

_____

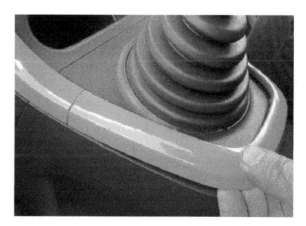

2 The above trim is typically fastened using plastic clips. Using a fine screwdriver the clip is pushed and the trim is allowed to pop up. Take care not to scratch the trim. If you can, unclip the trim by gently pulling it. If the trim is stuck look for other clips or screws – don't force it.

1 Check the door trim fastenings and gather the tools and equipment needed.

3 With the trim removed the necessary repairs can be completed.

4 The trim can now be simply clipped back into place.

On completion ensure finger marks are cleaned away.

How to remove a door trim

Enter your place of training or study and remove a door trim. Manufacturers use different methods but the picture sequence opposite can be used as a guide.

2 Locate the screws and clips that fasten the trim to the door, some may be hidden, don't force anything, work methodically and carefully.

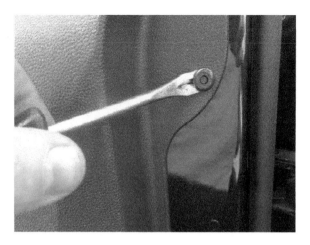

3 To release the above trim clip, the centre is pushed in, this allows the clip to be removed. These vary between manufacturers so take care.

4 Carefully pull the trim from the door, these are commonly held on with plastic clips.

5 Disconnect wiring for electric window switches and electric mirrors.

6 Before refitting the door trim ensure the waterproof membrane is in place and is not damaged.

Often screws that fasten the trim to the vehicle are hidden from view, look at the door trim below which has to be removed. The cover over the handle covers the screws and this clips on and can be easily damaged.

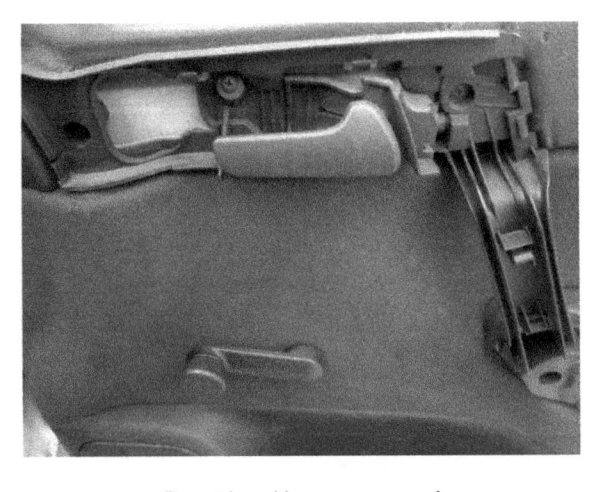

Door trim with cover removed

What is the best procedure if you accidently damage a customer vehicle trim?

Door trim with cover in place

 **TIP** After refitting trim components ensure they are clean and if necessary wipe them with a suitable cleaning product.

How to remove a wheel arch liner

Enter your place of training or study and remove a wheel arch liner. Safely raise the vehicle, often the wheel needs to be removed to gain access.

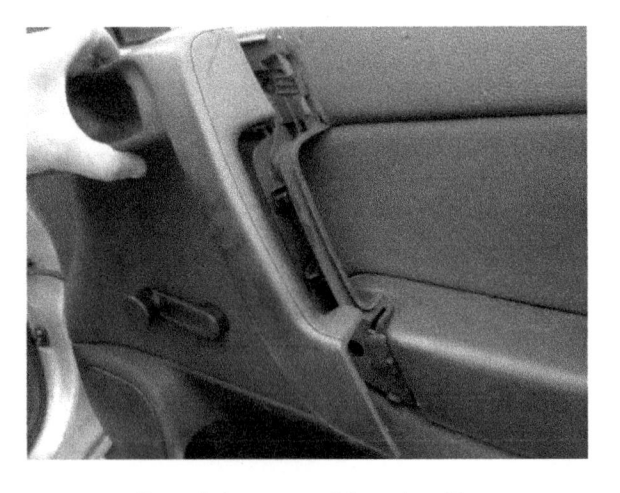

Cover being removed from door trim

Remove any fittings such as mud-flaps.

Locate the fixing screws or clips and remove,
taking care not to break them; often plastic expanding rivets,
screws or plastic nuts are used.

How to replace a wiper motor and linkage

It is common to remove a wiper motor and linkage. All manufacturers are different, enter your place of training and remove a wiper motor use this picture sequence as a guide.

1 Gather the tools and equipment to complete the task.

2 Remove the wiper arms, usually these are located onto splines. If they are difficult to remove find the correct puller for the job.

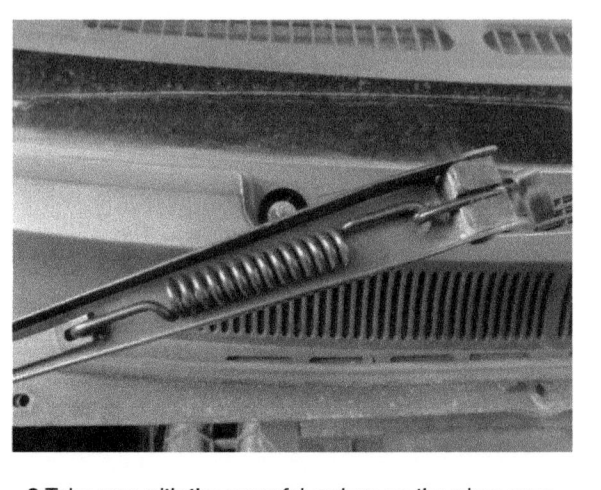

3 Take care with the powerful springs on the wiper arms.

Why are these powerful springs fitted?

_____

4 Locate the trim fixings; these could be expanding rivets, nuts, bolts or screws. Take care not to lose any.

5 Carefully remove the wiper trim, don't force it. If it is stuck, look for fastenings that you have not seen. Take care not to scratch the windscreen. Once removed, store the trim carefully.

6 Take care to disconnect the wiring. Ensure you don't damage fixings or connections, you don't want the wiring getting trapped in moving parts once the task is completed.

7 Remove the mounting bolts on the linkage and make a note of any earth wiring connections from the motor.

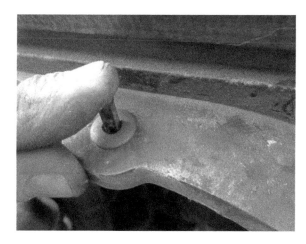

9 Ensure the trim is refitted without damage, refit all fittings and check everything is aligned and that water will be channelled from the windscreen correctly along the trim as it is designed.

8 Note the position of the spindle on the linkage, if necessary mark it with paint so that the new one is fitted onto the motor in the correct position.

10 Taking care run the wipers one cycle on intermittent wipe, this will park the wipers in the correct position. If possible use the dust mark on the windscreen to align the wipers. Tighten to the recommended specification.

11 Ensure that the wipers work correctly and they do not catch any of the trims, if needed re-adjust the wiper arms.

 Take care, wiper motors are very powerful. It is always recommended that you disconnect the battery first.

REVIEW

If you recover a vehicle that has been in an accident you may be required to quote for a repair, if sensors and components are damaged you must recognise them in order to complete the quote. See if you can recognise the parts below.

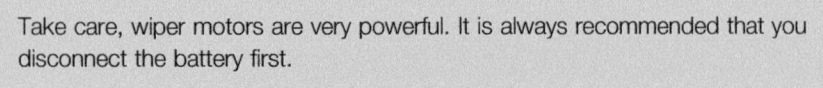

Complete a 500 word assignment on one of the following topics, relate this to your workshop experiences, add your own photographs and explain what the task involved, problems that were encountered and how the job was completed to a satisfactory standard.

- Mechanical components
- Electrical components
- Trim components

Multiple choice questions

Choose the correct answer from a), b) or c) and place a tick [✓] after your answer.

1 **A damaged vehicle has a steering column replaced, a sensor that may need calibrating is**

a) Steering angle sensor [ ]

b) Airbag deployment sensor [ ]

c) Steering tilt sensor [ ]

2 **How would you best describe a 'quote'?**

a) A near enough cost of repair [ ]

b) An approximate cost of repair [ ]

c) An exact cost of repair [ ]

3 **Technician A says that an airbag can be removed as soon as the battery is disconnected, Technician B says that you should wait a short time after disconnecting the battery, who is correct?**

a) Technician A [ ]

b) Technician B [ ]

c) Neither technician A nor B [ ]

4 **The next task once you have refitted steering and suspension components should be**

a) Road test the vehicle [ ]

b) Clean the car [ ]

c) Carry out alignment checks [ ]

5 **Before fitting a vinyl wrap it would be good practice to**

a) Repair damage and ensure the vehicle is clean [ ]

b) Spray the entire vehicle with a glue product [ ]

c) Mask off the windows and windscreen [ ]

GLOSSARY

ABS Anti-lock braking system.

Accumulator A reservoir to store liquid.

Actuator An electrical component that moves in response to a command e.g. a solenoid.

Adaptive suspension Suspension that can generally be changed during operation by an electronic control unit.

Annulus A ring gear with internal teeth which mesh with other gears.

Backlash The setting of free play between components.

Binding brake A brake that continues to work when the pedal is not pressed and the handbrake is released.

Blow-by Gases escaping past the piston rings.

Braking efficiency How well the brakes are working.

CAN Controller area network.

Capacitor An electronic component that can store a charge for a period of time.

Case hardened Heating steel to harden its surface and improve its strength properties.

Clearance A gap between two components.

Clutch drag A fault where the clutch does not release fully.

Clutch judder A fault causing unwanted vibration.

Clutch plate/centre plate/drive plate Turns the gearbox input shaft.

Clutch release bearing Thrust bearing that pushes on the diaphragm spring fingers.

Clutch slip A fault where torque transmission is reduced.

Coefficient of friction Ratio of the amount of force needed to cause an object to start to slide.

Continuing professional development (CPD) How people maintain their knowledge and skills related to their professional lives.

Corrosion A term used for rust.

Crown wheel Large gear wheel used in a differential.

Current The flow of electric charge through an electrical conductor.

Customer protection Legal rights of customers.

CVT Continuously Variable Transmission, which has an infinite number of gear ratios.

Damper A shock absorber.

Detent Used to ensure positive location of the selected gear.

Diagnosis Determining the fault.

Diaphragm spring A dish type spring used in pressure plates.

Differential Divides the transmitted torque equally to each output shaft and allows each to rotate at different rates.

Diode An electronic component that allows current to flow in one direction only.

Discrimination Being treated unfairly.

DVSA Driver and Vehicle Standards Agency.

Electronic control unit (ECU) An on-board vehicle system computer.

Epicyclic gear train A set of intermeshing gears consisting of the sun, planet and annulus.

Fabrication Building metal structures using cutting, bending, and assembling processes.

Fibre optics Data signals in the form of light are transmitted along thin glass or plastic fibres.

Four-wheel drive a system which allows engagement of drive to all four wheels.

Generator A device that can produce electricity.

Hazard Something that has the potential to cause harm or injury.

HCS High carbon steel.

Hybrid A vehicle that can operate using two sources of energy.

Hygrometer Used to measure moisture content.

Hygroscopic Absorbs moisture from the atmosphere.

Incandescent bulb A bulb that uses a wire or filament which heats up to give off light.

Inflammable Something that can be ignited or set alight.

Inhalation Breathing in chemicals and fumes.

Injector pulse width The time that a fuel injector stays open.

Interlock mechanism Used to prevent more than one gear being selected a time.

Intermittent fault A fault that may be present then not present.

Kinetic energy The amount of energy an object possess due to its motion.

LCS Low carbon steel.

Lean An engine that is running on a 'more air/less fuel' content.

Limited slip differential a device which limits and controls spin on each of its output shafts.

Misfiring Failure of an explosion to occur in one or more cylinders while the engine is running; can be continuous or intermittent.

MOST Media Orientated System Transportation.

Multiplexing A system which uses only two wires (CAN) for a number of ECUs to send and receive information along.

Oil consumption The use of oil by an engine.

Oil gallery A passageway in the engine which allows oil to flow to other areas of the engine.

Oscilloscope A device used to check for electrical signals/readings.

Overhauling Stripping down and repairing parts and components.

Oversteer The car turns more than desired by the driver.

PAS Power assisted steering, the system assists driver effort.

Passive suspension A suspension system which works as designed and cannot be adjusted for different driving styles.

Pinion A small gear on the end of the starter motor which meshes with the ring gear.

Pinion gear A gear which meshes with, and directly drives the crown wheel.

Pinking/ping A knocking noise caused by detonation (explosion) in the combustion chamber.

Piston crown The top of the piston.

PLTS Personal Learning and Thinking Skills. These are essential skills for work and general learning.

Potentiometer A variable resistor used to control voltage.

Pressure plate Bolted to the flywheel and presses the clutch plate to the flywheel face.

R134a Refrigerant used in modern mobile air-conditioning systems.

Rectification To repair something.

Rectifier The component that changes alternating current to direct current.

Refractometer Used to measure coolant, battery electrolyte and screen wash strength.

Regenerative braking Capturing the vehicle kinetic energy to generate electricity.

Release mechanism Means of the driver temporarily stopping clutch plate rotation.

Resistance The opposition to the flow of electrical current.

Reversible motor A motor that can operate in both directions by changing the polarity to the motor.

Rich An engine that is running on a 'more fuel/less air' content.

Risk The possibility of suffering harm, danger or loss.

Sensor An electrical component that changes its resistance or electrical current in response to the changes in a vehicle's characteristics e.g speed sensor.

Slave cylinder Hydraulic cylinder moving the clutch arm.

Solenoid A heavy duty electromagnetic switch which can be used in a number of applications, including the starter motor.

Spur gear A straight cut gear, which can easily slide in and out of mesh with a similar gear.

SRS Safety restraint system, this may include airbags and seat belts.

Stall test Is the method used for testing a torque converter coupling.

Third differential Used to prevent windup between the front and rear differential.

Three quarter floating Type of bearing arrangement used in vehicle hub assemblies.

Torque convertor A type of fluid coupling which multiplies torque.

Torque Turning force measured in Newton metres.

Torsion The action of twisting or turning.

Transistor An electronic component that acts like a switch.

Transponder An electronic device used to wirelessly transmit and receive electrical signals.

Understeer The car turns less than desired by the driver.

Vinyl wrap Vehicle covering used to advertise or change colour.

Viscous coupling A mechanical device which transfers rotation and torque by means of a viscous fluid.

Voltage The electrical potential difference between two points.

Warped disc A warped disc could be considered to be buckled.

Welfare A term used when discussing well-being of employees.

Wiper linkage The mechanical device that connects the wiper motor to the wipers.

INDEX

Abrasive wheel regulations 6
ABS *see* anti-lock braking systems
AC leakage test 281
accumulator 59
actuator 242, 243–7
adaptive suspension 144
air conditioning 295–6
    basic refrigerant circuit 297
    compressor 297
    condenser 298
    diagnosing faults 302–3
    evaporator 299
    expansion valve 298–9
    major components 297–9
    oils 299
    performance test 301
    receiver/dryer 298
    safety precautions 296
    service/charging ports 299
air conditioning servicing equipment 300
    charge 300
    dye detector 301
    electronic leak testing 301
    evacuate 300
    leak detecting 300
    recovery 300
air flow sensor 75–6
air suspension systems 149–50
    air springs 150
    compressor 151
    control module 150
    height sensors 151
    receiver drier 151

reservoir 151
    reservoir valve block 152
airbag suspension systems 152
airbags 126–8, 315–17
alternator faults testing/diagnosing
        279–80
    AC leakage test 281
    current output check 280
    noise checks 280
    voltage output checks 280
alternator wiring 279
annulus 208
anti-lock braking systems (ABS) 170–2, 305,
        318–19
    associated systems diagnosis 177–8
    brake assist 174–5
    diagnosis of sensors 172
    hall sensor operation 172
    modulator operation 173
    traction control 174
automatic gearbox
    automatic transmission 209
    continuously variable transmission 224–6
    control 226–7
    direct shift 215–16
    electric transmission control 229
    electronic control 222–3
    epicyclic (planetary) gear trains 216–18
    fluid flywheel 209–10
    gear selection 221–2
    Honda six-speed AMT case study 213–15
    hydraulic system 219–21
    mechanical system 218–19

oil cooler 221
    semi-automatic transmission 227–8
    symptoms, faults, causes 231–2
    testing and fault diagnosis 230–1
    torque converter 210–13, 230–1
auxiliary electrical systems 285

backlash 180, 181
bench testing starter motor 278
bent connecting rods 108
binding brake 158, 163
blow-dry 100
body kits 320
body repair 307–8
    customer estimates 311
    final adjustments 312–13
    health and safety 306–7
    mechanical components 311
    steering and suspension 311
    vehicle manufacture 308–12
booster 117
brake assist 174–5
brake drums and shoes 164–5
    measuring for 'ovality' (out of round) 165
brake fluid 160–1
brake master cylinder 186–8
braking
    anti-brakes and associated systems
        diagnosis 177–8
    anti-lock systems 170–5
    calculating efficiencies 167–8
    diagnosis and rectification of related faults
        159–60

drums and shoes 164–5
    efficiency 158, 166
    electronically controlled handbrake
        systems 168
    overhauling systems 185–8
    pads and discs 161–4
    regenerative 176
    roller brake tester 165–6
    systems 155–6
    vehicle stability control 175–6

camshaft sensor 79–80
camshafts 103
CAN *see* controller area network
capacitance 254
capacitor 250, 254
case hardened 44
catalytic converters 83–4
    diagnosis 86
center plate 192
central locking systems 293
    diagnosing faults 293
    smart keys 294
charging systems 278
    alternator wiring 279
    smart 282
    stop/start technology 282–3
claims 9
cleaning 20
clearance 100
clutch/es 193
    disc 193
    drag 192

clutch/es (continued)
faults and diagnosis 194–6
judder 192
plate 192
pressure plate 193–4
release braking 192
slave, cylinder, release bearing 194
slip 192
coefficient of friction 192
communication
interruptions 27
keep customer informed 27
listen carefully 27
negative questions 27
stress and emotion 27
technical knowledge 27
compliance bushes 145
compression test 64–5
compression tester 45–7
continuing professional development
(CPD) 23, 25
continuously variable (or stepless) transmission
(CVT) 208, 224–5
operation 226
Control of Substances Hazardous to Health
(COSHH) 4
controller area network (CAN) 250, 261
converter 116
coolant temperature sensor 77
cooling systems 97–8
corrosion 305
CPD see continuing professional
development
cranking current test 276–7
cranking voltage test 276
crankshaft end float 109–10
crankshaft removal and inspection 110
crankshaft sensor 79–80
crown wheel 233

current 250, 253
output check 280
customer protection 39
data protection 42
sale of goods act 42
warranties and guarantees 42–3
wear and tear 43
customer service 40
right first time 40–2
customers
angry/abusive 28
communication methods 27–8
giving body repair estimates 311
part worn tyres risk 13
CV joint boot removing/replacing 239–40
CVT see continuously variable transmission
cylinder bore finish 109
cylinder bore inspection 109
cylinder head removal 104–5
cylinder leakage test 65–6
cylinder power balance test 66–7

DAB radio 271–2
damper 144
data protection 42
dealership business 24
detent 197
diagnosis 39, 41
dice screwdriver
engineer's diagram 54–5
fabrication 53–4
diesel fuel injector faults 70
diesel particulate filters (DPF) 85–6
with additive 86
diagnosis 86
without additive 85
differential 233, 234–5
diagnostics (symptoms, faults, causes) 241
investigation 235

differential locks 203–4
diode 273, 278
direct petrol fuel injection systems 90–1
homogenous 91–2
stratified 91
direct shift gearbox (DSG) 215–16
discrimination 23
door trim 322–4
DPF see diesel particulate filters
drive line
differential 234–5
final drive and differential diagnostics 241
inter axle (third) differential 237
limited-slip differential 235–6
maintenance 238
removing/replacing CV joint boot 239–40
shafts and hubs 234, 238
symptoms, faults and causes 237–8
drive plate 192
drive train torque sensor 245
DSC see Dynamic Stability Control
DSG see direct shift gearbox
DVSA (Driver and Vehicle Safety Agency) 23, 25
Dynamic Stability Control (DSC) 175

earth circuit resistance test 277
ECU see electronic control unit
electric power steering 134–6
electric struts/dampers 148–9
electric vehicles 119–21
electrical components, common 313–14
airbags 315–17
anti-lock braking systems 318–19
wiper motors and linkages 317–18
electrically operated windows and mirrors see
mirrors; windows
electro-hydraulic power steering system
(EHPS) 131–3
hall sensor operation 133–4

electronic brake force distribution 173–4
electronic control automatic gearbox 222–3
operation 223
electronic control unit (ECU) 158, 242, 243
electronic principles 251
Electronic Stability Program (ESP) 175
electronic struts/dampers 148–9
electronic systems abbreviations/symbols 251
BCM (body control module) 251
CAN (controller area network) 251
CPU (central processing unit) 251
EOBD (European on board diagnostic) 251
IC (integrated circuit) 251
LCD (liquid crystal display) 251
LED (light emitting diode) 251
PCB (printed circuit board) 251
ROM (read only memory) 251
electronic transmission control 229
converter lock up 229
final-control elements 229
safety circuits 229
shift quality 229
shift-point control 229
employment
contract 33
discrimination 32
dismissal 33
rights/responsibilities 31–4
trade unions 33–4
engine management systems 75
engine related faults
knock sensors 62–4
noise 61–2
engine terms
boring 61
compression ratio 61
cylinder swept volume 61
energy efficiency 61
honing 61

mechanical efficiency 61
power 61
reaming 61
thermal efficiency 61
torque 61
valve lapping 61
Environment Protection Act 6
epicyclic gear 207
epicyclic (planetary) gear train 208, 216
compound 217–18
investigation 216–17
ESP see Electronic Stability Program
exhaust gas recirculation valves 80–1
extended range/plug in hybrids 118–19

fabrication 44, 52
exercise 53–4
hand tools 53
materials 53
tools and equipment 52–5
fault diagnosis 40
clutch 194–6
common rail diesel injection faults 70–2
diesel fuel injector 70
differential 241
drive line 237–8
electric windows 290
gather further information 40
intermittent faults 41–2
lighting circuits 266
manual gearboxes 199
petrol 68–70
repair fault 40
using experience 41
verify complaint 40
verify fault repair 41
fibre optics 242, 247
fire prevention 18

first aid 8
protecting yourself when helping others 8
fluid flywheel 209
assembly 210
four wheel steering 136
four-wheel drive (4WD or 4 × 4) 201, 202
fuel cell vehicles 121–2
fuel problems 68–70
future careers 25

gear selection (automatic transmission)
221–2
starter inhibitor 222
generator 113
good housekeeping
cleaning as you go 20–1
importance 17–18
poor 18–20

hall sensor operation 133–4
handbrake systems 169
hazard 2, 10
hazards 10
chemical 11
ergonomic 10
physical 10
recognising 10–11
HCS see high carbon steel
headlights 264–5
alignment 265
health and safety 3
air conditioning 296
blame and claim 9
body repair 306–7
common rail diesel 14–15
emergency first aid 8
hazards 10–11
high voltage vehicles 14

putting customers at risk with part worn
tyres 13
regulations 3–7
risk assessments 11–12
road testing vehicles 9
work related stress 12
health and safety terms and conditions 9
asbestosis 9
asphyxiation 10
carcinogen 10
competent person 9
COSHH 10
ergonomics 10
HASAWA 9
hazard 10
negligence 10
PAT testing 10
prohibition 10
RIDDOR 10
vibration white finger 10
heating 294–5
high carbon steel (HCS) 44, 54
high voltage batteries 115–16
high voltage vehicles 14
Honda six-speed automatic manual
transmission (AMT) 213–15
hybrid 113
hybrid vehicles 114–15
extended range/plug ins 118–19
high voltage batteries 115–16
invertor, converter, power boost assembly
116–17
motor/generator 118
transaxle 117
hydraulic lifters, replacing 107
hydraulic systems 128–9
hydraulic valve-control box 226–7
operation 227

hydro-pneumatic suspension 154
braking systems 155–6
pump and reservoir 154–5
suspension spheres 155
hydrogen vehicles 121
hygrometer 44, 50
hygroscope 158

incandescent bulb 305
included angle 140
independent garage business 24
inflammable 16
infotainment system 267
DAB radio 271–2
mobile communications 270
navigation 270–1
speakers 268–70
inhalation 16
injector pulse width 74
inter axle (third) differential 237
interlock mechanism 197
intermittent fault 39
intermittent wiper system 286
invertor 116

job roles 24–5
juddering 180

keys, smart 294
kinetic energy 208
knock sensors 62–4, 76–7

lack of performance 64–72
lambda sensors 82–3
diagnosis of 83
LCS see low carbon steel
lean 74, 84
learning through demonstration 36

learning/thinking skills 34–6
   creative thinker 34
   effective participation 34
   examples 35–6
   independent enquirers 35
   reflective learners 34
   self managers 35
   teamworkers 35
light emitting diode (LED) 263
lighting circuits 262–3, 266
   diagnostics (symptoms, faults, causes) 266
   headlights 264–6
limited-slip differential (LSD) 201, 235–6
liquid cooled alternator 281
low carbon steel (LCS) 44, 54
lubrication systems 94–7

magneto-rheological dampers 152–3
manual gearboxes 198–9
   faults 199
   testing and test equipment 199
map sensor 78
mass air flow meter 75–6
mechanical components 311
mechatronics 246
media orientated system transportation (MOST) 250
memory seat function 292
mirrors 290–2
misfire 59, 64, 67
mobile communications 270
modification additions 319
   spoilers and body kits 320
   vinyl wrapping 320
modulator operation 173
MOST see media orientated system transportation
motor/generator 118

multi-meter 48
multiplex 242, 246, 260–2

navigation systems 270–1
noise 61–2
   checks 280
Noise at Work regulations 7
NO$_x$ sensor and catalyst 84–5

O$_2$ sensor 81–3
oil consumption 93
oil cooler 221
oil gallery 93
oil pumps 95–7
oscilloscope 44, 45, 49–50
overhaul 180
   braking systems 185–8
   steering systems 181–4
   suspension systems 184–5
oversteer 158
oxygen sensors 81–3

pads and discs 162–3
   checking disc brake run-out 163
   measuring brake disc thickness 164
PAS see power assisted steering
passive suspension 144, 146–8
permanent magnet starters 275–6
personal hygiene 19–20
Personal Learning and Thinking Skills (PLTS) 23, 34
petrol 68–70
pinion 273
pinion gear 233
pinking/ping 59, 62
piston crown 100
piston protrusion 107
piston rings 108
plastigauge 110–11

PLTS see Personal Learning and Thinking Skills
potentiometer 284
power assisted steering (PAS) 125
   electro-hydraulic system 131–4
   fully electric 134–6
   hydraulic systems 128–9
   rack and pinion system 129–30
   steering gearbox system 130–1
power probe 49
power seats 292
   memory seat function 292
pressure plate 192, 194
pressure sensors 245

R134a 284
rack and pinion system 129–30
rail diesel injection systems
   diagnosing common faults 70–2
   health and safety 14–15
rain sensing wipers 287
receiver drier 151
rectification 74
rectifier 273, 278
refractometer 44, 51
regenerative braking 158, 176
regulations 3–7
relays 255
   delay 256
   testing 256
release mechanism 192
Reporting of Injuries, Diseases and Dangerous Occurrences Regulations (RIDDOR) (1995) 5
resistance 250, 253–4
reversible motor 284
rich 74
risk 2
risk assessment 11–12
road testing vehicles 9
roller brake tester 165–6

safety restraint system (SRS) 125, 126
Sale of Goods Act 42
screen heating 291–2
scrub radius 139–40
selector lever 246
self-levelling sensors 148
semi-automatic transmission 227–8
sensors 242, 257
   ABS 172
   air flow 75–6
   air suspension systems 151
   camshaft 79–80
   coolant temperature 77
   crankshaft 79–80
   drive train torque 245
   knock 62–4, 76–7
   lambda 82–3
   map 78
   oxygen 81–3
   pressure 245
   self-levelling 148
   speed 244–5
   suspension 147–8
   temperature 243
   thermistor 257–8
   travel 246
shafts and hubs 234
   faults and causes 237
   maintenance 238
   protection during use 238
skills, tools and equipment 45–51
slave cylinder 192, 194
slow speed wiper system 285–6
smart charging systems 282
solenoid 273, 274
solenoid valves 244
spark ignition engine 62
speakers 268–70
speed sensitive wipers 287

speed sensors 244–5
spoilers 320
spur gear 197
SRS *see* safety restraint system
stall test 208
starter generator 282
starter inhibitor 222
starter motor 274
 bench testing 278
 cranking current test 276–7
 cranking voltage test 276
 earth circuit resistance test 278
 operation 274–5
 starter relay by-pass test 278
 wiring 274
starter relay by-pass test 277
steering 311
steering axis inclination (SAI) 138
 included angle 140
 scrub radius 139–40
 thrust angle 141
 toe out on turns 141
steering gearbox system 130
 diagnosis 131
steering geometry 137
 axis inclination 138–40
 camber angle 138
 caster angle 137
 preliminary checks 141–2
steering systems overhaul 181–4
stethoscope 47
stop/start technology 90, 282
 operation of system 283
 starter-generator 282
stress, work related 12

suspension 311
 air suspension systems 149–52
 airbag systems 152
 hydro-pneumatic 154–6
 magneto-rheological dampers 152–4
 overhauling systems 184–5
 passive and adaptive 146–8
 purpose 145–6
 sensors 147–8
suspension terms 145
 bump 145
 ride height 145
 rotound 145
 semi-eliptical spring 145
 shimmy 145
 sprung weight 145
 un-sprung weight 145
switches 255

temperature sensor 243
terminology 126
test lamp 49
thermostat 97
 testing 97
third differential 201
three quarter floating 233
thrust angle 141
time keeping 30–1
timing belts and chains 102–3
titania type lambda sensor 82
toe out on turns 141
tools and equipment
 fabrication 52–5
 skills 45–51
torque 192, 194

torque converter 208, 210–12
 coupling characteristics 212
 lock up torque converter 212–13
 testing and fault diagnosis 230–1
torsen unit 206
torsion 125, 130
traction control 174
trade unions 33–4
transistor 250, 258–9
 acting as a switch 259
 operation 259
transponder 284
travel sensors 246
trim components 321
 removing door trim 322–4
 removing interior trim 321–2
 removing wheel arch liner 324–5
 replacing wiper motor and linkage 325–8
turbo chargers 88–9
 inspection 90
two-speed transfer box 202–3

UEGO (Universal Exhaust Gas Oxygen) type
 lambda sensors 82
understeer 158

valve inspection 104–5
valve stem oil seals 106–7
variable valve timing (VVT) 86–7
vehicle Air Conditioning regulations 6–7
vehicle manufacture, body design 308–11
vehicle stability control (VSC) 175–6
vinyl wrap 305, 320
viscous coupling (VC) 201, 204
 centre differentiated 207

and differential 206
 operation 205–6
 torsen unit 206
voltage 250, 253
 output checks 280
VVT *see* variable valve timing

warped disc 180
warranties and guarantees 42–3
washer systems 288
waste disposal 21
water pump 98
waveforms 252–3
welfare 16
wheel arch liner removal 324–5
wiggle test 39
windows 288–9
 bounce back 290
 diagnosing faults on electrically
  operated 290
 one touch operation 289
windscreen wiper systems 285
 intermittent 286
 rail sensing 287
 slow speed operation 285–6
 speed sensitive 287
 washer systems 288
wiper motors/linkages 284, 317–18
 replacing 325–8
workplace structures 24–5
worn tyres 13

zirconia type lambda sensor 82